PENGUIN BOOKS

CAPTAIN COOK, THE SEAMEN'S SEAMAN

Four-masted barques and full-rigged ships were moored within sight of the house in Melbourne where Alan Villiers was born. From his earliest youth he was determined not only to sail in [illegible] ut to command one. When he was fifteen he [illegible] the Tasman Sea Barque *Rothesay* [illegible] ury he wandered the se[illegible] ab dhows. He became p[illegible] en de Cloux; master ow[illegible] d ship to sail round the [illegible]. and Squadron-Comma[illegible] n Sicily onwards through [illegible]. In his ship *Joseph Conrad* he saile[illegible] the *Endeavour*, and he has been a lifelong stu[illegible] of the works of the great seaman, Captain Cook. He has sailed in old-time square-rigged ships probably as much as any man alive – even back to the seventeenth century in the *Mayflower* replica. He is a trustee of the National Maritime Museum, Chairman of the Society for Nautical Research, and on the Board of Governors of the Cutty Sark Preservation Society. Alan Villiers is author of more than a dozen books on ships and the sea. Many have been translated into several languages, and *Cruise of the Conrad* and *The Set of the Sails* are recognized sea-classics.

CAPTAIN COOK, THE SEAMEN'S SEAMAN

A Study of the Great Discoverer

Alan Villiers

ILLUSTRATED BY ADRIAN SMALL

Master Mariner in Sail

PENGUIN BOOKS

Penguin Books Ltd, Harmondsworth, Middlesex England
Penguin Books Australia Ltd, Ringwood, Victoria, Australia

First published by Hodder & Stoughton 1967
Published in Penguin Books 1969
Reprinted 1970

Made and printed in Great Britain
by Hazell Watson & Viney Ltd
Aylesbury, Bucks
Set in Linotype Granjon

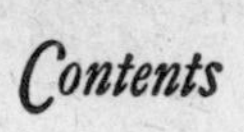

Contents

Maps

Preface

A PRINT on the wall of the classroom at my school when a child in Melbourne showed a benevolent Captain Cook landing from H.M.S. *Endeavour* at Botany Bay. He was restraining some of his men from firing at a handful of Aborigines who stood there in courageous and futile opposition, with their spears and woomeras. I liked the look of Captain Cook. So I read all I could of him. I liked what he had done and the way he did it. When I could, I went as an apprentice in Tasman Sea barques – a reasonable equivalent to the Whitby 'cats' of the eighteenth century, offering remarkably similar training. I became a professional sailing-ship seaman, as he had done. I sailed where he sailed, in similar ships descended in long and then unbroken line – doing the world's work. In time I became a Master, and sailed a full-rigged ship not much different from his *Endeavour* round the world in as much of Cook's tracks as I dared. It would be difficult for a ship to sail, say, from the Panama Canal to Sydney even on a great circle course and not cross and recross Cook's Pacific tracks at least a dozen times. But I do not mean that sort of thing. My *Joseph Conrad* wandered the Pacific in the Roaring Forties and the Shrieking Fifties: I made wide 'swings' in the Trade wind zones: I beat past the ghastly Great Barrier Reef, out of the Coral Sea. I touched upon a reef there (for an abundance remains) and got off the way Cook did by carrying out anchors with my boats and heaving off by capstan operated by the brawn of my crew. I rounded Good Hope and the Horn, because that was the square-rigged way. I had sprung spars; I lost a t'gallantmast; I dodged ice; I hove-to in violent gales, sparred with Tuamotuan reefs and South Seas calms. I looked for anchorages round the Sandwich Islands and found them scarce.

Then and later, I have tried to go where Cook went, from

Unalaska and Cook Inlet to the Antarctic, from Botany Bay to Batavia, Green Island to the desolate grandeur of South Georgia, from Nootka Sound to Newfoundland and Quebec. *All* sea discovery was by sailing-ship, conducted by sailing-ship seamen. I count it my personal good fortune that I became a seaman while the tradition of the deepwater engine-less square-rigged ship was still unbroken and an accepted way of sea life. I may well be the last man on earth who has sailed a ship like Cook's round the world with the power of the free wind, taking and using what comes. And so I may see perhaps a little more clearly than most just how great and infinitely admirable were the achievements of Captain Cook: perhaps, too, I may help to present him and his work in a manner comprehensible when all deepsea sails have disappeared from the face of the sea for ever.

For Cook to me is no visionary who stumbled upon what was there because it *was* there waiting for the first stumbler to find it, nor an exploiting late-comer reaping the harvest of predecessors' long and painful pioneering. He was (and he still is) the meticulous and infinitely careful explorer by sea, the most consistent and the greatest sailing-ship seaman there ever was.

ALAN VILLIERS

Oxford

CHAPTER ONE

A Whitby Seaman

THE slowly freshening soft spring wind hummed a low song in the sails and the small ship's rigging, and the bluff old bow pushed the tideway waters gently aside with the thrust of the broad, straight stem. Along the sides Thames water talked back quietly as the hull slipped by, and the coal-laden Whitby 'cat' *Friendship* came beating in from sea. Aft the water reflected in the stern windows set the scintillant light to dancing there, from the bubbling, busy wake. It was not a grumbling, reluctant wake nor a fussy one, for the hull had homely, sea-kindly lines to slip through the sea with minimum protest. Father Thames knew this ship well and all her hard-working kind, for they were the north-country colliers, the three- or four-hundred-ton workhorses which brought the coal of which the port of London was for ever in need. A collier the *Friendship* certainly was, but there was no speck of coaldust on her trim and spotless decks or aloft on the sails shining in the sun. Here everything was trim and shipshape, as in good ships it has to be.

The wind was ahead, coming down the river from the west. The ship leaned to it as it freshened slightly. The array of blocks and cordage by which her sails were kept in control creaked efficiently as it took the strain, and then yielded, as it must, and the *Friendship* picked up more speed. To wind'ard and to leeward of her, upstream and downstream, the whole wide estuary was alive with plying ships, inward-bound, with here and there a big outward-bounder, stemming the tide fussily, standing out to sea under a gracious spread of wind-filled sails.

The tall seaman at the wheel was alone on deck. A striking man, young, over six feet tall and built to scale, he seemed as much part of the ship as her three masts were. From the big jib set from the boom-end for'ard to the small spanker on the mizzen aft, the old barque-rigged cat sailed in perfect balance.

On the wind, deep-loaded, she sailed herself, for long usage had developed perfection in the set of a square-rigger's sails when trimmed and tended by those who understood them. The tall seaman, adjusting the wheel with a spoke or two to keep her taut on the wind, smiled at her handiness, for he felt in harmony with his ship as a sailor should, or a horseman does with a splended mount, perfectly under his control and responsive to his slightest gesture however communicated.

Slipping the spliced eyes of a couple of small lines over two spokes of the wheel to hold the helm in place while the ship took care of herself, the Mate walked quickly for'ard to peer beneath the foot of the foresail and to leeward of its leach, for the sail obscured his vision from aft. These were congested and dangerous waters, but a swift look at the well-known landmarks of the Kent shore reassured him. The mains'l was hauled up to have the ship more manageable, and the spritsail – that perfect blinder out on the bowsprit – was fast. Leadlines lay at the ready for soundings in the fore channels on both sides, but he knew where he was. For the moment they could wait. He eased the bowlines and saw that all the braces were taut, for he meant to shake her properly for a few moments to take her forward way off and let the flood tide push her bodily upstream against the wind. He knew the ship was now in the tide's strongest grip. Further towards the Kent shore it weakened and there were sands. He knew those shoals and banks as well as he knew the ship's gear. He was aware, too, that if she did touch on the flood she should come off again. At the worst and she got stuck, she would sit down on the sand with her comfortable, almost flat-bottomed hull, like a duck on a nest, and come to no harm. It was not true to say that she had been designed to do that. Coal cats and coastal sailers generally were not designed. They had been evolved, 'grown up' in the trade they served, slowly through trial and error and the painful survival (if fortunate) of its perils, becoming through the centuries perfectly suited to it. He lifted the beckets on the wheel, brought the ship slightly to the wind, letting the square sails shake. First the weather clews of the t'gall'nts began to lift, very slightly: then the leaches trembled and began to shake, and the shaking continued down the weather sides of all the square sails as if the ship were trying to shake the wind out of her as a spaniel, fresh from the water, shakes off the spray. Her forward speed dropped to a knot or so. She stayed almost in the same water, rushing up-Thames towards London more or less sideways.

This was exactly what was needed as the barque made a long, slow reach across the broad river and the Mate on deck, skilfully tending her, gave everybody else – both watches – time to scoff their midday meal down below. They ate fast with determination and table manners sufficient to the purpose, but it took time for the cook to dish up each man and boy's mess of robust stew and the 'afters'. By alternately giving the barque her head for a few moments and then letting her shake and drift sideways in the tide, a coasting Mate could give the crew time to eat between the goings about at the end of each tack. This was known as a 'collier's nip', for the colliers were in the river most often and were most expertly handled.

The manoeuvre was not as simple as the tall Mate of the *Friendship* made it appear. The river was full of shipping all under way, all highly individual, the inbound hurrying for their discharging berths, the outbound big fellows bleary-eyed with weeks in port, their crews not yet efficient, many of them probably still drunk and both crews' and ships' behaviour therefore unpredictable.

The year was 1755. There was not yet any defined and internationally binding Rule of the Road for ships, but by long custom many usages and practices were established and rigidly followed. Each time he shook her sails the Mate well knew that the barque, for the moment, was not properly under control and, to that extent, could menace other vessels. There were at least fifty other deep-laden colliers beating up, crossing and recrossing the river in a wonderful pageant of wind-blown shipping covering the best part of a century, a pattern of symmetrical beauty where the sun touched all those sails that, though they took it for granted, stirred the tough salt-soaked hearts of all who saw it. Sometimes, plying* their separate ways, they passed within feet of each other, each with confi-

* Plying: beating, tacking.

dence in the alertness, the competence, and the skill of the other.

One thing all those plying, deep-loaded square-riggers had in common was that, by the nature of the angle of their awkward sails to the wind, they would not respond to the helm very quickly. The sails had to be handled, too, if they were to fall off from the wind: the lee main braces let go as the helm was put up to push the ship from the wind, the spanker sheet let go to take off the leverage aft; or, if she were to come into the wind, fore and jib sheets must be eased to let her swing her head up unimpeded. It all took a nicety of judgement, a sure touch of long-acquired skill that had become second nature through years of practice, a clear head and strong, competent hands, perfectly co-ordinated. Slight variations in the local wind or vagaries in the swift-running tide could affect ships differently, though they were close together, and, at a moment's notice, change safe passing to a dangerous collision course. None of them would go swiftly off the wind: in such cases both must put their helms hard a-lee to come into the wind, for then if they did touch it would be a brushing together at the worst and not a smash-up collision. A ship with one man on deck, no matter how skilful, obviously had to take very special care to keep out of all others' way, especially as unwritten law required her to go about at the edge of the tide, even if the tide ran in only half the channel. Aboard the *Friendship* only two men were trusted to be alone on deck, in charge of the ship, in these conditions – the Master and the Mate. Captain Ellerton was not much older than his Mate, James Cook, but he had been much longer at sea. A Whitby shipmaster began early and retired early too, if he could. First he had to live frugally over many years and save money hard to acquire a little capital, and so buy some sixty-fourth shares in well-run ships. The sixty-four divisions in ship-ownership were traditional, too, like the ships themselves. At sea everything was traditional. Men stayed in the sort of ships they knew. When they could retire ashore, they stayed with these ships if they could, too, in management. It was more than a good tradition. It was the eighteenth-century way of life.

While he tended the barque in the crowded tideway – a process so familiar to him that it did not occupy all his mind – the Mate found himself thinking of all this seafaring life and his own part in it. He thought of his own career – a rather odd one, in some ways. In those days when youth was apprenticed to the sea almost in childhood and irrevocably bound, at latest, at the age of twelve or thirteen, he had begun at eighteen. Why so late? Why to sea at all? He himself hardly knew, for it was a harsh profession – dangerous, too. Yet there was *something*, some spirit of ships and the sea which he had found compelling, since first seeing big ships in the North Sea from the hole-in-the-cliffs hamlet of Staithes, on the Yorkshire coast. His first sight of a score or more in harbour at Whitby, close-to, with sailors working in their rigging, had been breath-taking, as he looked down from the hill-side by the river. The sturdy but graceful ships, the white clouds of their sails drying from their yards, the bowsprits stretching for the sea, stirred his imagination as nothing had before. Here was a man's life and he knew it was for him.

He was already seventeen, nearly eighteen. Apprenticeship ended legally at the age of twenty-one. He was a farm-worker's son, born inland (at a hamlet called Marton-in-Cleveland, in October, 1728). Noting his good mathematical mind, his Scots father had sought what education he could for his bright son. Others had helped. It was a slight schooling but better than most. Farmhand James Cook senior, by long application, by dint of his own good head for figures and sturdy worth, rose to be manager of a neighbour's farm. He was anxious that the young James might do even better. It was no part of his plans that the lad go to sea. So he was sent to the village of Staithes to learn the business of running a general store which concentrated on plain grocer's goods and plainer drapery. Dull stuff and a dull career – but what offered? In the eighteenth century, a farmhand's son was lucky to acquire a brief working basis of the three 'R's' – Reading, Writing, and 'Rithmetic – and even luckier to escape the endless drudgery of working on the land. Once permitted to enter the fringes of that world of 'trade', only the shiftless and the indolent would turn from *that*. Go to sea? The man before the mast followed a hereditary calling forced on him by economic stress and lack of other opportunity. Once off to sea, he clung to his ship stubbornly with pride and an abiding sense of comradeship and mutual confidence among shipmates of his kind. Adventurous? Yes, perhaps at times – too much so: but the most probable adventure was early death, from drowning if coastwise, from disease if deepsea, with the prospect of a hot 'press' into the Navy to hasten both ends. For merchant seamen did not voluntarily enter King George's naval service, and in time of war the press was aimed primarily at them. What possible inducement was there to accept such a life voluntarily, when it could be avoided?

Yet seamen also could know glory. Now and again throughout history they had risen to the heights with their fragile ships, in battle or on dreadful, questing voyages. There was an indefinable magnetism, a sort of charm, for adventurous lads in the sight of tall masts, gracefully sparred, and the song of the wind in the rigging. Staithes was near enough to Yorkshire and Northumberland ports to have its quota of old sea-dogs whose yarns were stirrring and infectious. Whitby itself was then one

At Whitby

of the principal ports of England. There were Whitby men who had sailed to the Indies, Whitby men who had gone – and were still going – on whaling voyages among the Arctic ice, Whitby men who knew Greenland's nearer waters as well as they knew their own grey old North Sea. Whitby men, sailors said, had been with the buccaneer-discoverer William Dampier on his voyage with privateer Woodes Rogers in 1708 to 1711, when they rescued the marooned Alexander Selkirk, the prototype Crusoe.

Yarns of such men and their adventures real or imaginary were heady stuff, but the service of Whitby's North Sea ships was another thing. Apprenticeship to any calling was a serious matter in the eighteenth century, above all to ships and the sea. Owners were reluctant to accept a youth well past the normal age, for obviously they would get fewer years of work for nothing out of him, and the mariners were equally loath in their anxiety to maintain the standards of their arduous profession and, indeed, just normal workaday pigheadedness. What, said the able seamen, a young fellow learn all they knew in only *three* years? They had taken seven, or more: this was the proper thing. An apprentice out of his time had to face searching examination by a committee of elderly able seamen before being admitted to their ranks, and they were rigorous. Men and officers alike never forgot that at sea the safety of all could depend at any moment on the alertness and competence of any. A man rated able seaman had to be just that, and no shade of doubt about it.

It took a year or more of special pleading and persuasion, plus the obvious unusual quality in Cook's own clear eyes and the cut of his countenance, to get the youth to sea at all. The perspicacity of a couple of elderly Quaker shipowning brothers – the Walkers, of Grape Lane, Whitby (where their old house still stands) – got him off at last in their coal-carrying barque *Freelove*. Here he was one of half a dozen apprentices. The *Freelove* was 450 tons, built at Whitby, a 'cat' * like so many

* The term 'cat' as a ship's description referred to hull shape and not to rig: the differentiation of ships by standardized rigs came later, in the nineteenth century. A 'cat' was a stumpy carrier with rather Dutch or

others, rigged as a simple barque without either refinements or frills. Her business was to earn money by carrying goods, and coal was the cargo offering most readily – coal from the nearby north-east ports to London, sometimes across to Norway or the Baltic. Occasionally, if one of the too frequent nondescript wars was on – usually with France – she might be chartered as a navy or army transport for coastwise or North Sea passages with stores and cavalry horses. Whitby cats had capacious holds and were good transports, and reliable.

It was the usual practice in the coal trade to lie-up at Whitby during the worst of the winter months, from a day or two before Christmas until the first good weather in early March. During these times the ships were rigged down, their spars sent to spar sheds, their rigging to rigging lofts, their sails to the sailmakers, while the hulls were fumigated and whatever defects could be found on deck and below and in their simple 'machinery' – a main deck capstan, a for'ard windlass, two catheads and cat tackles for getting the anchors stowed (common to all sailing-ships and nothing to do with the name 'cat'), simple wheel and tackle steering – were put in order. Whitby shipwrights did the woodwork: Whitby crews did the rest. Apprentices were lodged ashore, in attics of the owners' or Master's homes, while their ships were laid up, but they worked from dawn to dusk. It was a full life, but at any rate while ashore they were dry and well fed.

It was also the ideal life for learning the sailing-ship seaman's business in all its practical aspects, just as the North Sea, with its hazards of sandbanks and racing tides, lee shores, poorly marked roadsteads and indifferent harbours difficult to approach,

---

Norwegian lines – apple-cheeked for'ard but fine aft, deep in the hull, flat-floored to sit on the bottom and stand up with minimum ballast. Such ships had often to sail back to their loading ports empty. Obviously the less ballast they had to buy the more economical they were. Chief characteristics of the cat to outward view were very bluff bows above the waterline, a straight somewhat unlovely cutwater without any form of figurehead, and a slim, five-windowed stern. The two hawseholes cut in the flat 'face' may have given them a sort of feline appearance: there seems no other reason for their name of cat, though it could have been derived from Ancient Norse and have nothing to do with four-legged cats at all.

frequent gales and sudden shifts of wind, provided an excellent and most testing sailing ground. No wonder mates and seamen were strict and zealous of the standards of their calling!

An apprentice was expected to be first aloft (if he could beat the men), first with his handspike to the windlass or his bar to the capstan, quick on deck at the call for all hands – a frequent summons – and ready for cuff or clout if he were last, and the dirty jobs if he were lubberly. The aim was that he should not just know his work but excel at it and exult – in time – in that excellence. Pride in their work was such a characteristic of these collier seamen that many carried their own handspikes with them the better to attend the windlass: Masters said the anchor came in twice as fast if there were Whitby men aboard. In ports on voyages, the boys were kept aboard, with minimal leave to stretch their legs, for the temptations of the waterfront were abundant, appealing, and obvious. There was also, very often, the risk of losing them to the naval press-gangs. Apprentices under eighteen were supposed to be immune, if not put aboard by the parish, but the press needed seamen above all others and had to find them. As for the parish boys, their treatment was a typical piece of eighteenth-century meanness. Poor children, brought up at the expense of the parish, were got rid off as quickly as possible, the boys often bound to the sea at twelve or thirteen. Having been reared at the community's expense, they were held to be available for the community's defence in time of danger. As for the merchantmen, they could find more such boys and train them, for there was always a good supply of parishes willing to be rid of their indigent young.

Even the parochial authorities had reservations about offering the boys directly to the Navy, which had no real facilities for training them. Lads from the colliers were active and competent aloft and on deck, and made excellent top-men. If a cannon-ball ended their existence, why, it was quicker than drowning. After every on-shore gale much of the east coast of England – especially the shoal-encumbered bulge of Norfolk and Suffolk, thrusting out into the sea as a trap for ships unable to claw off the land and, uncertain of their positions by night or in the sea haze, unable to run for harbour either – was cluttered with the bodies of seamen young and old and the wrecks of ships. For the whole of the way from 'the Spurn to the Thames the channel through which ships must pass is between sand banks and the main, and frequently between one sand bank and another'.* There were few lights to guide the mariners, and those few were mostly coal fires burning fitfully in ill-tended braziers on some windy headland. Taylor declared that ships had 'to lay-to and fire guns to awaken the drowsy attenders, and oblige them to stir up their fires'.

Through many of the 'gatts' – channels – they would not sail by night at all for fear of the fatal embrace of some ghastly sand: for how could they know exactly where they were? Use of the lead was constant from both fore-channels, the brawnier apprentices swinging the heavy sea lead there day and night until their whole bodies were soaked from the wet leadline and their chanting voices telling the depths became a pathetic croak. The coasting coal-cats sailed by the three 'L's', which for them were Lead, Lookout and Local Knowledge, equally important.

Mr Cook thought about these things. For nine years now, in four ships over a dozen voyages (not reckoning those frequent coal hauls down to London) he had served as apprentice, able seaman, and for the last three years, as Mate. It was a hard life even when you liked it. He remembered one late autumn voyage, for instance, from Shields with coal for London. Fifteen ships sailed: by the following morning only six were left. The others were all overwhelmed by a storm which blew up suddenly soon after they had crossed the bar. A strong northerly wind increased to a gale with the coming of night, so strong that all sail had to be taken in but the fores'l and a rag of a trysail, aft. Under this, with the tack down to the weather cathead and the yards braced, they stood to sea to get an offing, for the master feared that the wind would turn to N.E. or E.N.E. If he ran

* *Memoirs of Henry Taylor From 1780 to 1821* (North Shields, Printed for the Author, 1821). Taylor was born in Whitby in 1737 and went to sea at thirteen, in Whitby ships. He commanded ships in the coal trade – 'that best nursery for seamen' – for twelve years.

before it he would get his ship in such a position that soon she could not clear the land at all. Wind and sea would force her inshore while the land itself, bulging outwards, got increasingly in her way – the old, fatal problem of North Sea coastal sailing.

All night the wind increased and the sea with it. Rolling heavily, the small barque shipped a lot of water. The sea took everything movable overboard, led by the only boat. In the blackness of the gale-filled night the barque became quite unmanageable. The trysail blew out. They could neither get the fores'l in nor set a little more sail to help balance her. The noise of the gale and the crashing seas was terrifying even to men well used to it. Young Cook had known many a wild gale flinging its salty assaults on Staithes, but to be in a small ship tying to live through such a thing was altogether different. No friendly light blinked through the night to help indicate their position. The gale worked more easterly. The fores'l blew out with a great booming like thunder and a thrashing of split canvas, writhing and tearing at the rigging as if determined, like a mad thing alive, to destroy all it could. Seamen ran to the jeers to lower the yard, but it would not come down. The sail thrashed the wilder.

'Man the gear! Aloft and furl!' came the orders, shouted above the storm.

Up climbed the sailors. Now without a single sail, the ship fell in the trough of the sea. Here she rolled and rolled in such an endless agony of motion that it was all her sailors could do just to hang on. The wind forced them against the rigging, flattening them as the ship fell down to leeward, crushing the breath from their lungs; then as she picked herself up and, with a wild toss of her head, flung herself back to wind'ard, tried to throw them bodily outboard, far into the sea. Yet they still fought, though the flayed ends of stout sail cloth tore writhing at their eyes and the gale became a personal enemy, powerful and horrible, determined to destroy them.

So daylight came. They were off Flamborough Head, about twenty-five miles more or less south of Whitby and sixty-odd from Shields, six or seven miles to seaward. They thanked God for that, for they still had an offing. Inside Flamborough Head the coast trended westward for a while. There was an anchorage in Bridlington Roads with at least some shelter from winds north of east. If their anchors would hold they might ride out the storm there – if they could get in.

But the attempt was useless. The heavy-laden barque would not steer. She lay in the trough of the sea bashing herself to pieces in the sea, tried almost beyond endurance, her motion like that of the pendulum of a huge old clock which the furies were playing hand-ball with.

Even the youngest apprentice could see that there was no hope of making Bridlington Roads.

Around them were other hard-pressed vessels fighting for their lives. One of these, a fair-sized ship belonging to Vickerman's in Scarborough, was very close. Apprentice Cook could see her people manning both the pumps. Her sails were in ribbons, her decks gutted, the ship deep in the water, obviously sinking. She was driving out of control towards the Whitby barque. Collision was imminent. The larger ship drove fast: the smaller could not get out of the way. Perhaps the Scarborough captain, as a last desperate act, hoped to touch the ships together while his crew leapt across for their lives – a desperate hope indeed for if the ships touched in that sea and smashed and ground together, it would be brief encounter fatal to both.

The Whitby captain, seeing this, while his crew looked in horror, braced his yards for'ard in order that their windage and that of the half-blown-away sails stowed ill upon them might give his barque manageable steerage way, if only for a moment.

It worked! She drew slowly ahead. The other – rolling, lurching, her three mast-heads gyrating wildly in the screaming sky as if beseeching the grace of the Almighty in last hope of succour – just missed their quarter.

It was her last act. One moment the sea flung her high as if lifting her up for one final look at the savage universe: then, as the crest passed, dropped her down, down and drowned for ever, with all her people. Nobody jumped. Neither ship had a boat. Neither had anything floatable left on deck. The ugly sea, snarling over the spot where the ship had been, looked for the moment almost satisfied.

The situation in the Whitby barque was still desperate, though she was sound and leaked little. Another long night the gale blew, but it worked back to the north-east. On the second morning they could set the mainsail and, by the exercise of magnificent seamanship, work their vessel behind Spurn Head and let the anchors go. All hands had been awake and hard at work for three days and nights. The sight of their fellow-seamen's obliteration was harrowing. They let go the main jeers, too utterly worn out to haul the sail up and secure it, and fell asleep in their sodden clothes as they stood.

When the wind eased and hauled to the north again, they sailed on for London. The passage took a month.

There were good passages, of course. Dangers were accepted and dealt with, if possible, as they arose. Such was the life. Now he, James Cook, who had embarked to the sea voluntarily when he had other prospect, stood at a cross-roads. Messrs Walker of Whitby, well satisfied with him at the age of twenty-seven, had offered him command at the end of this voyage. If he accepted he could be established for life. Walkers were good owners. *If* he accepted? He should be greatly pleased, he knew. And yet. ... Command of a Whitby collier-cat, fighting the wretched North Sea voyage after voyage for the sake of a few coal-cargoes brought to London – was this the end of all ambition? With maybe a passage every now and again to Norway or into the Baltic, or a spell on coastal and cross-Channel trooping when there was a war. A career at sea was sharply defined in those days, circumscribed for the lowly by the manner in which it was begun. Even a Master was confined to his own familiar trade, knowledge of which was a hard-won thing, personal to himself.

Yet the seas were wide. He looked again at this pageant of shipping on London's river, at a big East Indiaman going down slowly in a skilled pilot's care. (For the exalted officers who took such ships on their passages to India disdained the humdrum work of getting down the river and did not come aboard until the ship was wholly ready for sea.) She was beautiful, a superb thoroughbred of a ship, graceful of line and symmetrical rigging, her sides pierced for guns as a warship's were. Watching her carefully, giving her a wide berth, Cook passed close enough to hear the farmyard sound aboard, the cows lowing and the crowing of cocks, and the shrilling of pipes, the strange sound of some mellow Asian gong. She was indeed a fantastic ship, and she offered a great life while it lasted.

But not for him – *never* for him nor the likes of him. James Cook like other working seamen well knew that recruitment for command or other senior, profitable posts aboard Indiamen was by family influence, wealth, and preferment ashore. Competence hardly came into it. Command of an Indiaman could be the road to real wealth, and bright young men from England's exclusive schools, finding themselves without other preferment, vied for the officer berths aboard. Such employment brought the chance of fortune, for all officers had rights to ventures of their own from which the profits could be enormous. Captains sometimes made thousands of golden sovereigns on one voyage, retired ashore rich men after three or four.

Mate Cook knew also that even if he could get such a berth, he could not afford the money to rig himself out as the lowliest of an Indiaman's petty officers. And then what? The style of life, the profusion of men and hope of profit – no, somehow these things did not appeal to him, even if available.

His eyes swept the decks of his trim cat – no fighter she, except against the sea, and no great profit-seeker either. Coal-laden and lowly, condemned to coasting and the narrow seas, her service leading in the end to command and then, with luck, slow rise to petty ownership – a few sixty-fourth shares first, zealously nurtured by good management and the utmost economy to grow into shares in other ships and, perhaps, slowly to competence enough to move ashore and progress into lesser capitalist.

There was something lacking in this prospect, now open to him: a vague but very real restlessness had hold of him, not for the first time. Perhaps his seafaring until now had been limited, a coastwise floundering on soundings mainly, when the seas were so wide.

His eye caught another sort of vessel, a great three-decker coming down from the Chatham dockyard – imperious, stately, powerful, and beckoning. The sun caught her gilded figurehead, shone upon the opened blood-red lids of her scores of gunports, turned her high masts and maze of rigging to a flood-lit wonder. She moved with grace and perfect form through the swift river waters, her hull and masts and sails blended into a striking whole which was at once romantically adventurous, powerful, and profoundly efficient. From her no farmyard sounds arose: her stance upon the water was assured and thoroughly harmonious – a challenging ship, a challenging career.

James Cook stared at her. *Here* was a ship. And with her, a life that really could be satisfying, that could bring some answers to a man's questing.

'There'll be a hot press for that one, Jamie,' broke in the quiet voice of Captain Ellerton, emerging from the aft companion, his lunch over. 'This war will grow, mark my words. She'll need everyone she can get. Best stop aboard tonight when we get to our berth.'

His Mate's eyes followed the ship-of-war. Slowly he turned and faced the Master like a man who had at last arrived at a firm decision, long thought over.

'I'll not wait for the press,' he said, very deliberately. 'I've a mind to try that life. I'll volunteer.'

Captain Ellerton looked dumbfounded. 'Why, Jamie, you're next for command! You've got immunity in the Whitby ships as long as London needs coal or the Army needs transports. *Volunteer* for the Navy? Why? You'll make Master, maybe, but you'll get no further. Mark my words! And you'll be allowed no thinking for yourself, ever.'

But James Cook, thinking deeply, made no answer.

'We're coming to the edge of tide. Tumble up the lads! 'Bout ship! We'll put her 'round. Hard-a-lee!'

In a flash the barque spun in her stride. Sheets slatted. Sails thrashed briefly as they expelled the wind from one side, swung, and took it on the other. Blocks sang. Yards swung. Hemp rattled happily as braces, bowlines, tacks, sheets did their work swiftly and efficiently in expert hands. In a trice the cat was round.

As she beat on up the river her tall Mate looked back at the trim three-decker, and there was a settled, thoughtful expression on his open face as if he knew at last what he meant to do.

CHAPTER TWO

Master, R.N.

Cook's decision to enter the Royal Navy as a volunteer seaman, though astonishing to his shipmates and other Whitby seamen, was no sudden change of mind, nor was it taken lightly. He knew that, as a working seaman without influence or well-placed friends, Captain Ellerton's remark that he would make Master but nothing else, summed up his prospects. He knew also that there was room in the Navy for a practical seaman who knew his business, and the rank of Master offered an honourable career. The Master's position was a sort of anachronism, a hang-over from the days when soldiers and their officers fought at sea as necessary but did not handle ships, nor learn much about them. The military officers fought with their men, and the ship's Master sailed the ship as the senior military man aboard wished. A Master served his ship, maintained and sailed her, and commanded his crew of seamen. His career and his calling was 'low-caste': leading fighting-men was a 'high-caste' business, and always had been. Therefore Masters had no social pretensions or standing, which meant among other things that recruitment for them was from *working* seamen, career professional men, mainly from merchant shipping. There was no aristocratic competition. Down the centuries the status of the Master had remained constant. He was the senior deck-working officer in the merchant service and the King's ships alike, in charge of maintaining the delicate maze of the rigging and the suits of sails, *sailing* the ship where she was required to go, giving those complicated verbal orders which remained for ever incomprehensible to all landsmen, but were the evolved and essential essence of the sailing-ship's being.

He was also, usually, the pilot and the navigator – different callings, for the pilotage of ships then was by eye and personal knowledge and navigation was, in part, by careful astronomical observations with precise instruments long handled, and the most careful reckoning. Good course-keeping, judgement of leeway, accurate estimation of speed under sail – for there were no adequate instruments to measure or record it and, the wind being fickle, the sailing-ship's forward speed varied infinitely – and the assured ability to appraise performance of his ship in any conditions, all came into this. In the Navy, the ubiquitous and omnipresent Master was charged also with the survey and recording of roadsteads, bays, anchorages, ports, and rivers where his ship touched. He had to see the ship's log was properly kept and write his own journal, recording those matters of professional concern to the ship such as the wind and its variations, the weather, the courses made good, positions of the ship as known by bearings and distances off headlands, and all that sort of thing.

To add to his responsibilities, as if these were not enough, he had charge of the ship's cleanliness and her stores. He worked her people and her sails. He supervised her topmen. He tended her at exposed anchorages, and was responsible for her clear hawse and clear cables, which is to say that he must see that, when an anchor was let go, the long hempen cable ran out clear and did not snarl itself in a dangerous heap of knots, bends, and twisted snags: and, when the ship was safely anchored, that she did not drag, or swing in such a way that the cable loosened the anchor's hold instead of steadying it, or ride over her own anchor at slack water and then as the tide dropped, sit down upon its pointed flukes – a destructive process which could be fatal. In short, the Master of a King's ship was Master indeed, a career man with a job to do, and his hands full doing it.

To Cook, who had seen something of the Navy when on coastal and Irish Sea convoys in Whitby colliers taken up as transports, the service offered prospects and a way of life which for him were preferable to the dead-end business of coastwise coal-hauling. The Master of a coal cat, though a first-class seaman, was a foreman at best, and his Mate a leading-hand. The Mate's hope lay in promotion to Master: the Master's was advancement by the slow accumulation of capital and assiduous attention to the business side of his ship even more than the seafaring. This was the road to ownership. The business side did not appeal to Cook: if he'd wanted that he could have stayed with the storekeeper of Staithes.

With a light heart, as soon as the *Friendship* was berthed and her coal swinging dustily out, Cook took himself to the Thames-side village of Wapping where the Navy had a depot. Here he entered as able seaman – the only rating open to him. It was a beautiful June day in 1755, and the recruiters at Wapping were anxious to get hold of all the seamen they could. The Navy was building-up at speed. There was no 'declared' war on at the moment, but warlike preparations increased daily – against the French, as usual. The British government's policy was to crush French power and put an end to French colonizing on the North American continent, where they were doing all too well. If the French could not be defeated in Europe – there were far too many of them there, and they had a habit of producing good armies and able generals – then their power must be thwarted. They would have to be contained in America, and India too: thrown out of both, if possible, and kept out. For this it was necessary to attack them boldly in Canada, to dominate the Indian Ocean by the effective use of sea-power, and to maintain a blockade against France.

It was nearly a century and a half since the British had established themselves in Virginia and New England, but progress there was slow – the young America was still little but a string of loosely allied and sometimes quarrelsome colonies along the Atlantic seaboard, their backs to the great American hinterland, their eyes on Europe. Behind them, French adventurers from Canada sailing across the Great Lakes and marching south, had come upon the Mississippi and the boundless fertile plains through which it flowed. English settlers, groping through western Pennsylvania and upper New York State, had reached the Lakes, too. American Indians raided the remoter areas of New England too often and too successfully, and there was more than a suspicion that France was behind them. It was the colonial habit then to scorn and suppress the Indians: the French treated them better. English and French clashed, inevitably. If French penetration had continued, the wars of the

Revolution might have been against France, not England: or might never have been thought necessary at all.

The new seaman-recruit was sent forthwith to Portsmouth to join the 60-gun ship *Eagle*, Captain Hamer. On the way down in the lumbering coach (which rolled and pitched far more than the *Friendship* did) passengers talked of an army under the new Commander-in-Chief in America – one General Edward Braddock, sixty-year-old veteran Scots warrior, a guardsman with plenty of fighting experience in Europe but none at all in America, where he had then been for a few months only – who would soon chase the French out of there. Braddock was to be reinforced from England, they said, and a fleet was assembling at Plymouth for the purpose.

The coach was full of businessmen, more interested in the prospects of making money from outfitting the new fleet than in the outcome of any war, declared or undeclared. Cook sat on the small bundle of his sea clothes – all he had – and said nothing. He watched the lovely fields of southern England go slowly past the coach windows as the vehicle lurched painfully along the rutted road, and thought of his new career.

The *Eagle* was in a mess. Everything needed doing and nobody was doing it. To James Cook, she was a big ship – four times the size of the *Friendship*. Her yellow sides with their blood-red open gunport lids towered above the fitting-out berth. Her high-steeved jib-boom seemed as long as the whole *Friendship*: her masts and lower yards were enormous. The numbers of thick shrouds and hempen backstays necessary to support the masts, and the size of the blocks and multiplicity of gear to control the sails and yards were bewildering. But Cook, reporting himself to the grizzled old Master in active charge, soon sorted out the apparent maze, for the *Eagle*'s rigging was only the *Friendship*'s grown up. He fell to with a will, soon found himself quietly leading the small group of skilled seamen setting up the head-gear, and they found themselves as quietly accepting him. In square-rigged ships, seamen's work was a skilled business carried on in the public view, and fellow seamen could quickly appraise the quality of a newcomer.

So could Masters, and First Lieutenants, and Captains.

Within a few weeks Cook was made Master's Mate – *one* of the Master's Mates, to be accurate. A Master's Mate was a petty officer. In a big ship, the Master had as many Mates as the establishment allowed and he could find worthy of the job, for his own responsibilities, Cook discovered, were even more extensive and onerous than he had thought. A King's ship was strictly for fighting. She was – or had to become as quickly as possible – a highly efficient gun-platform, propelled through the water by effective sails set upon and from perfect masts and yards and rigging. Since she had no cargo, she must be properly ballasted and in perfect trim.

Since merchant seamen rarely volunteered and there was no other source of experienced personnel, there was always a shortage of good men. To make up for the primitive conditions of life aboard the grog flowed freely. Beer was the staple drink and the allowance was a gallon a man a day, with half a pint of rum or brandy instead when the beer ran out. The beer was often poor, so the seamen mixed it with spirits to make a potent drink they called 'flip'. The rum and brandy were not diluted: that refinement came later. In Cook's day, spirits were served straight. So was tobacco – from the purser, at a price. The allowance was two pounds a man a month. It was strong, coarse stuff which the seamen smoked in clay pipes, or chewed. A man who could habitually chew sea tobacco did not notice the foul taste of his salt 'beef' rations, for he soon had no taste at all.

The *Eagle*, under Captain Hamer, was put when ready to blockade duty on the beat from Land's End to Cape Clear, in Ireland. This took in the Scilly Islands, treacherous in the days of sail. It could be a stormy area with the westerly gales of the North Atlantic blowing fiercely home on a rockbound lee shore, after having the whole Atlantic to whip the sea to fury. Patrol is worse than voyage-making: life in the *Eagle* was tough. In at least one aspect it was far worse than any Cook had known in Whitby ships. This was in the important matter of health. After a few months on patrol the *Eagle* landed 130 men seriously ill with scurvy. She had buried twenty-two more at sea, including her surgeon – all this while almost in daily sight of land. This sort of thing had never happened in Whitby ships which kept the sea only long enough to deliver their cargoes, and were not crowded. A King's ship had to have men enough to serve and fight her guns, plus replacements, as well as topmen, men to sail her, marines, and all the rest. The only places for them to live were the gun-decks, always draughty and frequently wet. A man had fourteen inches to sling his hammock, and no more: headroom was sufficient for dwarfs. Food was terrible and cooks were worse, for the chief requirement in selecting them was that they should be useless in other employment and, above all, never possess all four limbs. If they had two arms and two legs they could fight. One arm and one leg were enough for a cook. Food was stuff which could be stowed in barrels and kept – salted meats from inferior beasts, ancient cheeses of poor quality full of red worms, 'biscuit' full of weevils, pease and oatmeal full of anything but nutrition, a little rancid stuff which might once have been butter. This was the lot. The issue was adequate while it lasted, but the value was slight and, while days at sea lengthened, steadily worsened.

As for sanitation, the officers had their personal night-commodes and the quarter-galleries aft where they could crouch in comfort – they had their own food too, and usually drier quarters – but the men could relieve themselves only over the side. Right for'ard beyond the main body of the ship, at and around the heel of the bowsprit was the traditional place for them. It was also the wettest, windiest and most dangerous place in the ship, and the men crouched desperately there in rows like wet scarecrows, clinging on for their lives, hanging to their trousers with one hand and grabbing for handhold with the other, while the spray drove over them as the ship dipped and plunged, and threatened to throw the lot of them in the sea. This was the 'heads', not to be confused with figureheads. The whole forepart of the ship, where her bowsprit grew and her outwater clove through the sea, was her 'head'. At least it was well washed, for it was constantly wet if a breeze blew at all and, to that extent and no other, perhaps hygienic. There was literally nothing which the old sea-dogs could do in comfort, not even die. Death from wounds was barbarous and from scurvy lingering, loathsome and miserable.

Cook noted much that was new to him, and wondered that seamen, so hard to train and to recruit, should at sea be considered so expendable.

How long he might have been with Captain Hamer in the *Eagle* without making much progress in his new profession, no

one can say. But Hamer was relieved of his command for some dilatoriness in refitting after a storm when Cook had been a few months aboard, and the new captain's name was Palliser – Captain Hugh Palliser. Here a wise Providence stepped in on Cook's behalf, for this Captain Palliser was an officer of exceptional ability, discernment and drive, destined to become an admiral of the White, a governor of Newfoundland, and a Lord of Admiralty. He also became the influential patron of James Cook throughout his service life. Under Palliser the *Eagle* saw action and plenty of tough cruising off the Bay of Biscay and the Western Approaches, much of it in winter. The tall Yorkshire seaman, with his striking figure, his strong north-country face, and his very obvious competence soon came to the new captain's notice. Within a month or two he was made bos'n – real promotion, for a Master's Mate could be just another dog's-body about the decks while a bos'n was an important petty officer in his own right.

A 60-gun ship like the *Eagle* was not hazarded too close to the nasty coasts of Europe on her stormy 'beat'. It was usual to give her a group of smaller vessels to help, cutters and the like, or to rig and sail her boats. These spread over the sea to extend the warship's vision. Bos'n Cook soon had command of one of these, his first command. She was a 40-foot sloop with one minute cabin and a few small guns.

If life in the *Eagle* was tough, in the cutter it was just survival. Independent cruises, more or less in touch with the mother-ship, lasted two or three weeks. Cook's first was made in the spring of 1756. His log from the period survives, a laconic, briefly factual document kept 'on B^d ye Crewzer Cutter', much sea-stained and weather-beaten, with occasional entries about giving 'chace to a Sail to ye E^d', or 'Brot too & exam^d a Spainish Snow Bound for Bilbow' (stopped to examine a Spanish brig sailing for Bilbao), but mainly about wind and weather and the position of the little vessel. It was useful experience: it showed Palliser, too, that his bos'n did not need the backing of the discipline of a big ship to exercise power of command.

Back aboard the *Eagle*, she was soon in action as better weather tempted French shipping to slip out from their ports. Palliser sailed after two of them, took both, and put his bos'n aboard the larger to sail her to London. She had a full cargo of sugar and coffee which would bring good prices there, and prize money was important. In the meantime the *Eagle* was at Plymouth to land sick – scores of them, all with scurvy – and replenish men and stores. At sea again, crusing with the *Medway* in company, the two ships gave chase to a big Frenchman off Ushant. All night they shadowed her, closing in. Dawn showed her to be a 1,500-tonner, a big French East Indiaman converted to a ship-of-war, with fifty guns, her gunports open and the 18-pounders run out, ready for action.

Typically, Palliser was ready too, with his crew at action stations, decks sanded, surgeon's quarters all ready and the saws sharpened, powder-monkeys on the job, and all other preparations made during the night. He sailed right in to the fight though the dilatory *Medway* was hove-to and stopped, while her people cleared her for action. Within a couple of ship's lengths 'close and very brisk fire on both sides' lasted for the best part of an hour, when the *Duc d'Aquitaine* surrendered. She had recently landed a very rich cargo from the East Indies at Lisbon, which was unfortunate for her captors. The *Eagle* was lucky to have only ten dead and thirty-two seriously wounded, but both her own and the Frenchman's rigging was shot to pieces. Cook was busy supervising a jury rig for the *Eagle* to stagger back to Plymouth. A prize crew from the undamaged *Medway* looked after the Frenchman: this was the other ship's only contribution. Of the *Eagle*'s war complement of 493 men, fifty were dead and over a hundred seriously wonded. She was almost totally dismasted. Wind and sea got up while the hulks rolled and pitched in the surly Ushant waters, at the mercy of the strong tides. Tired from the long night's chase, worked up by the stress of the sharp action and perturbed, perhaps, by the sight of their shipmates cut down and mangled, the bos'n's party got on with the difficult and dangerous technicalities of setting sail enough to keep their ship under effective control and off the French coast.

Back in Plymouth, it was obvious that the *Eagle* needed a considerable refit. Her crew were paid off. About this time a letter from the local member of Parliament for Whitby reached Palliser, through Admiralty, in which the grant of a commission was requested for Cook. The captain had thought of this himself, but the regulations – not always observed – required that Cook (or anybody else from the lower deck) should have six year's service before such promotion was possible. Whether the politician's letter had much influence or not, shortly afterwards Cook was stepped up from Bos'n to the warrant rank of Master. On his twenty-ninth birthday, 27 October 1757, he was appointed to the *Pembroke*, a 64-gun ship, destined for considerable active service on the North American station. He had been some two and a half years in the Navy. Taking into account also his sound North Sea training, he was an experienced and competent seaman. He had shown himself to be a fearless, sober, and thoroughly dependable petty officer, well worth a master's warrant. But he had never been off soundings, nor required to use his skill at deepsea navigation. If there had been no more to him than the abilities and skills he had acquired up to that time, it is unlikely that the world would have heard of James Cook, and Australia would probably be French.

But the Master of the *Pembroke* became a noted man in the American campaign. He was in luck. William Pitt was directing the war for England, and Anson had been in dynamic charge of Admiralty long enough for his reforms and organizing skills to permeate through a navy much in need of them. Pitt was determined to bring security to the English colonies in America. For this it was necessary to curb the French there once and for all. His strategy included a three-pronged attack into the heart of Canada – one prong advancing through the British colonies taking the French Fort Duquesne (now Pittsburg) on the way, another up the Hudson valley and so (in time) to Quebec, the third up the St Lawrence to assault Quebec directly. Of these the last was most important: its success was vital. Early in 1758 a strong fleet under Admiral Boscawen took 14,000 of Pitt's best troops to Halifax, with forty-one-year-old Colonel Jeffrey Amherst commanding them. Halifax was back-base. The fleet took more than three months to sail the comparatively short (but rarely easy) passage from Plymouth to Nova Scotia, which must have surprised James Cook in the *Pembroke*. It was his first trans-ocean passage: he never made a worse. It took so long for the soldiers and sailors to recover from the inevitable scurvy that the troops could achieve little that year, and the other 'prongs' did no better.

One of Amherst's brigadiers was an officer named Wolfe, a sickly young man – he had, among other serious complaints, tuberculosis of the kidneys, with gravel and rheumatism as well – but an extremely able general. For a long time the war (known in history as the Seven Years' War) dragged through the usual stupidities, errors, rivalries of politicians and others, and all those aggravating, unnecessary and destructive tragedies great and small which seem inseparable from wars. Pitt's bold idea of a direct assault on Quebec was audacious and had, at first, every chance of quick success. But it bogged down. The French, firmly established on what seemed to be an impregnable height beside a dangerous river which was thought to be unnavigable for large ships, considered themselves secure. The St Lawrence was a notoriously fast and difficult river even for a small sailing-ship to navigate: the French had never attempted to reach Quebec with ships-of-the-line. Now the British boldly planned to sail a fleet there by night, and take the stronghold by frontal assault.

But time dragged on and on. Delays of one sort and another held up the campaign. All 1758 dragged by: early in 1759, as soon as breaking ice made the St Lawrence navigable for anyone, some ships of war did reach Quebec. But they were French. The British admiral left to keep guard from the base at Halifax was still there: nobody had told him the ice was moving out, and he hadn't looked. The enterprising French officer who stole a march on him was Bougainville – by odd chance the same who was to make a great Pacific voyage a few years later, and almost beat Cook to Australia – and, as well as food and military stores, he brought an intercepted letter in which Amherst outlined the whole plan of campaign. The alerted French had ample time to make Quebec all but impregnable, and built up a force of 16,000 under their able general Montcalm. Wolfe had 9,000. The French, very naturally, also removed all the marks in the St Lawrence which made the river usable. This they reckoned would put an end to any hope of the planned amphibious assault.

But they reckoned without the maritime ability of the aroused English, at last on their mettle. They reckoned without the courage and determination of thirty-two-year-old James Wolfe. They knew nothing whatever of the extraordinary abilities of another thirty-two-year-old, a lowly Master in the English fleet, by name James Cook.

For the English had to learn quickly how to sail safely up the St Lawrence and do what had never been done before, bring the largest ships of the time within gunshot of Quebec's forts. A major part in this desperate and difficult business was entrusted to the master of the *Pembroke*. His admiral was Sir Charles Saunders, one of Anson's men: like Palliser (and the shipowning Walkers back in Whitby) Saunders knew a seaman when he met one. Night after night for a week, Cook led the other masters in sounding, examining, buoying the channel. One extremely difficult place known as the Traverse almost defeated them.

Here the river so filled itself with awkward navigational hazards that the devil himself could not have done better: ships had to twist and turn like powerful twin-screwed liners, though they had only their sails, dependent on such wind as blew, and the strong downstream set was always against them. Rocks and shoals surrounded them: and they must sail by night. If the English remembered it (and the strategists certainly did) there had been a previous occasion to deter them. In 1711 a large English expedition tried to sail up the St Lawrence to Quebec – 5,000 troops in transports, convoyed by a score ships-of-war. None of them saw Quebec. The reefs and the rocks defeated them: so many ships were wrecked and soldiers drowned that the survivors quietly withdrew. The memory stuck, as it does with failures.

Sharp-eyed Indians with wonderful night vision saw the ship's boats, in which the Masters were stealthily sounding, and joined the French attacking them. There is a story that one night Cook narrowly escaped a strong Indian attack, leaping out of his boat at the bow as a dozen scalp-hunters yelling war cries jumped into the stern. But buoying, sounding, observing useful natural marks, recording odd sets in the racing river, the work was done – and done so well that, afterwards, Cook converted the data then so perilously gathered into a chart of the St Lawrence so accurate that it was not superseded for more than a century.

At last – at long last – all was ready. On a black night in September 1759, the fleet moved up. No ship grounded. No ship swung out of control, broadside on to the fatal rocks. The alarmed French sent fire-ships against them, did all they could.

But the fleet came on inexorably. James Wolfe was able to lead his men up the heights to the Plains of Abraham, and himself to death – and deathless fame among the military immortals.

Knowing of this victory the Master of the *Pembroke* slept soundly, for the first time in weeks.

Cook had been four years in the Navy, two years a Master. One reads of him being consulted about landing-places and the like by Wolfe himself. Though Masters were respected as competent seamen, they were in no sense expected to be real surveyors, and Cook could have no prior local knowledge. It is obvious that he could have reached the inner councils only on the recommendation of his admiral. It is equally obvious that he did not suddenly excel. After this, he is found employed as Master of Lord Colville's flagship, H.M.S. *Northumberland*, still on the American station – studying in the long winters, surveying and charting spring, summer and autumn, turning in charts of a standard hitherto unknown. Lord Colville had especially asked for him as an outstanding man: indeed, he was already regarded as the best pilot – master of the science of pilotage, which is the safe conduct of ships in narrow waters despite all hazards – in the Navy.

How could the Yorkshire farm-labourer's son have reached such a standard so quickly? A short answer is that no one else tried. There were plenty of good Masters, but a pint of heady 'flip' for most was a better and more human diversion than a surveyor's open boat. They did their jobs and they did them well, but no more. Even so early in his career, Cook was noted among his seniors as the man always ready to do more. Soon he had his own command, the little schooner *Grenville* – a New England tops'l schooner, a handy, saucy-hulled, 68-tonner previously named the *Sally*, built in Massachusetts in 1754 and bought into the Navy for £327 15s. od. – which was used for his surveys. It was as well the *Grenville* ex-*Sally* was a handy vessel, for the Newfoundland coasts Cook had to chart are among the toughest in the world. The cold Labrador current bringing its ice, broken pack and big 'bergs down from the Arctic past those iron-bound cliffs towards the warmer waters of the Gulf Stream Drift is a fog- and storm-breeder of unique ability. Old-timers said Newfoundland had three seasons only, July, August and winter. From the seaman's point of view Labrador had only one – winter.

But a 70-ton schooner on independent service could be a satisfying command, despite the difficulties which, after all, were not so much to a man brought up in Whitby ships in the North Sea. Cook – still a Master only – was now also 'Mr Cook, Engineer, Surveyor of Newfoundland and Labrador', cartographer and noted mathematician as well. He had been studious since early youth. The Walkers' housekeeper at Whitby, who looked after the apprentices when their ships were laid up, singled him out for his own 'table and candle, that he might read for himself', and fetched a second candle before the first spluttered out.

He was a youth then. As a man, his studious habits did not leave him. There is a memorable word-picture of him left by Major Samuel Holland which describes how Cook, Holland himself, and Captain Simcoe of the *Pembroke* worked together in the great cabin of that ship in July, 1758, when the ship was at anchor off Louisburg. Holland, a professional military surveyor, had been called in for some urgent special work, and Cook watched him at it, with the greatest interest. Cook was especially interested in the use of the Plane Table and quickly learned to use it, and, for a while, became Holland's assistant. When the *Pembroke* was at Halifax waiting for the long winter to be over, instruction continued in her great cabin.

'Whenever I could get a moment from my duty, I was on board the *Pembroke*,' wrote Major Holland, 'where the great cabin dedicated to scientific purposes and mostly taken up with a drawing table, furnished no room for idlers. Under Capt. Simcoe's eye, Mr Cook and myself compiled materials for a chart of the Gulf and River of St Lawrence. ...' Holland was surveyor-general of Quebec from 1764 to 1801, and surveyed the area for settlement by loyalists after the American Revolution. Cook had to learn from military surveyors because there were then no naval: Captain Simcoe was an unusual officer.

One likes the scene in the *Pembroke's* great cabin, that

spacious and well-proportioned room copiously lit by its line of bright stern windows – an apartment infinitely more gracious and attractive than any found at sea today, with evidence of its practical association with war at sea in the shape of large ringbolts here and there on the uprights, to anchor the tackles of extra guns, and the black guns themselves for the moment quiescent. Major Holland and Mr Cook get on with the enormous business of rendering an infinity of trigonometrical data and calculations into charts for the use of seamen, while benevolent old Simcoe now and again gives sage advice, and help also.

And so Cook learned. He was no man to neglect opportunities, and both Captain Simcoe and Major Holland were able men. According to the major, Captain Simcoe 'recommended him (Cook) to make himself competent by learning Spherical Trigonometry with the practical part of Astronomy, at the same time giving him Leadbitter's works' – *A Compleat System of Astronomy*, and *The Young Mathematician's Companion* of 1739 – 'of which Mr Cook, assisted by Captain Simcoe's explanations of difficult passages, made infinite use...'

This was during the winter of 1758–9. Later, after the Seven Years' War, Cook was surveying from the schooner *Grenville* for five more years, sailing the vessel back to England to work up his charts and sailing directions each winter – tough going for a 70-ton schooner, for the winter North Atlantic can be as nasty as the seas off Cape Horn. It was invaluable experience, the more worth-while as Sir Hugh Palliser was governor of Newfoundland for much of this time and kept a benevolent eye on his unusual protegé. Cook was still Master only, but as a surveyor his pay was ten shillings a day, which was more than most lieutenants earned. There was now peace. Many lieutenants were unemployed, on 'half-pay' they rarely received. Cook's money was assured and constant. He earned it.

He was indeed indefatigable. On an August day in 1766 there was an eclipse of the sun. The weather was fine, for once. He was on the south-western side of Newfoundland not far from its extremity there, Cape Ray. So he made a series of very useful observations during the eclipse, while carrying on his survey. When he had time, he worked these observations up, and sent the resultant paper to that fountain-head of scientific knowledge, the Royal Society in London. Here it proved most useful and is, indeed, still preserved. Learned gentlemen of the Royal Society heard of James Cook for the first time. They were to hear again – often.

By one of those coincidences which life so often provides, Cook encountered – but did not then meet – an illustrious member of that same august Society a month or two later, at the crater harbour of St John's. Cook was in port preparing the schooner for her homeward passage. At anchor close by was the big H.M.S. *Niger*, in which one of the officers was a lieutenant named C. J. Phipps,* also preparing for homeward passage. It was 27 October: Cook sailed on the following day. That evening, a boat pulled round the schooner's counter, coming from the *Niger*. In the crowd seated on beflagged cushions in the stern among the handsome, gaily uniformed officers sat a striking young gentleman not in uniform, an obvious civilian, most gaily dressed of all. The boat passed within yards. The young gentleman looked up with an infectious smile.

'I wonder who that young fellow may be,' said Cook to his Number One.

'Mr Banks,' was the reply. 'Mr Joseph Banks, the naturalist. He has been gathering specimens in Newfoundland and Labrador. I heard that Mr Banks and Lieutenant Phipps were students at Oxford together.'

Cook looked at the bright young man with added interest. Oxford? He had heard of the university: that was about all. But influential young graduates with friends able to get passages for them in H.M. ships, bright young men of obvious substance, who wanted to botanize and holiday in odd spots like mosquito-ridden Newfoundland and Labrador – why, such were new. They were indeed rare, not only at St John's.

Joseph Banks, Esq., F.R.S., was going to the Governor's ball.

* This was Constantine John Phipps, later the second Baron Mulgrave, with whom the youthful Lord Nelson was to sail a few years later on a voyage in search of the North-West Passage, as a very young midshipman. Captain Phipps was at Oxford with Banks, joining the Navy later than most officers. An earlier kinsman had been a colonial governor of Massachusetts.

The governor was a good friend to Mr Cook, but he did not invite Masters to social occasions.

Mr Banks had in fact nearly died of a fever ashore in Newfoundland. If he had died, not only James Cook's life might have been different. For James Cook, described several years earlier by Lord Colville as a seaman 'of genius and capacity' well fitted for 'greater undertakings', though still known by name only to experts and a few officers with unusual powers of perception, was soon to know a change indeed. In this, he was to see a great deal of the adventurous Mr Banks.

CHAPTER THREE

Misnamed Pacific

As the earth is spheroid, if you travel far enough going either east or west, you will be back where you started. The proposition sounds simple now. Yet the earth was accepted as a sphere two thousand years before anybody managed to travel round it. You had to sail. You couldn't walk or ride. You had to have seaworthy ships and good seamen. Given these, the difficulties were far from over. No nation was given these. They had to be evolved. The trouble with globe-circling was two-fold – first the Americas, for so long unknown; then the Pacific. That odd and difficult visionary Columbus was thwarted by the Americas though he thought he had sailed to the Far East. Since the days of ancient Rome it was appreciated that the best manner to sail from Europe to Asia should be by sailing west. What Columbus and Magellan really discovered was that there was too much America and too much Pacific in the way. The northern extremities of America were lost in impenetrable ice and no ship could ever reach the Pacific that way, during the sailing-ship era. The southern extremity, reaching far down towards the Antarctic, was beset by dreadful gales, the preponderance of which blew directly in the face of any ship trying to find her way towards the west. There *were* passages at either end – two in the south – but the northern was strictly for ice-breakers and regular use of the southern had to await the coming of first Californian clippers and then large steel sailing-ships in the nineteenth century.

The Pacific Ocean covers an area greater than all the land in the world. It occupies a third of the whole globe. From Arctic to Antarctic, from the west coast of the Americas to the east coast of Asia it straddles the world. Before it could be used by sailing-ships its wind and current systems had to be understood, for these are the life-blood of sailing-ships' voyages. The place seemed almost designed to keep Europe's sailing-ships away, though a coastwise Asian commerce had existed there from time immemorial. This used the defined limits of favouring monsoons which were early understood. Most such sailing was on the western side of the Indian Ocean, and in the China and Java Seas – mainly not in the Pacific at all. The Pacific was left to blow-aways and fleeing islanders who fled one way and did not return, migrating onwards in the tropic belt of the vast Pacific where the trade winds blew. Thence, having developed large double canoes which could carry victuals enough and endure a while in tropic seas, at least a few made voyages which were not just island-hopping – from somewhere in the Society Islands or Tahiti westwards and south to New Zealand, or north, with the trade winds abeam, to Hawaii. Chinese junks were stout ships and made long voyages in Eastern seas; but though they sailed long distances this was done by coasting, in favourable seasons, well understood. Arabs, Indians, Malays did likewise, the Arabs perhaps over a much longer period and on longer regular voyages than the others.

But these all knew Afro-Asian waters only, and the most able Asian seamen, the Japanese, did not venture at all. Japanese ships could have had fair winds to sail towards Canada and northern California, north of the tropics where the westerlies blow, and favouring winds back again with the Pacific's north-east Trades. They had a beam wind down to Australia – a sailing passage which was not much more than island-hopping for them. But they did not go. No Japanese, no Chinese, no Arab, no Asiatic knew how to sail in the Pacific.

For this there was good reason: the place offered no known trade and their own countries – the ports they understood – offered all the trade they could handle with (at least much of the time) the political stability to make trading voyages possible. Japanese seamen were forbidden to leave home waters.

Why then venture into the unknown?

By medieval times at least, Asians had ships good enough for such long passages. Chinese junks and Arab dhows were excellent ships a thousand years ago. Their seamen knew the compass. They had developed nautical astronomy. They could navigate quite adequately when necessary, and safely make such open-sea passages as from East African ports to south Indian. (It was an Arab pilot who showed the Portuguese Vasco da Gama the way from Malindi to Calicut.) But, in good-weather seasons, it was easier to coast.

In the Pacific you could not coast.

It was possible for Malay seamen to cross by prau to tropical north Australia from the nearer Indonesian islands, to gather bêche-de-mer and such. They were not traders but gatherers of something they could bring back for the traders to buy. No one could induce the Australian aborigine to engage in trade. They were gatherers, too, and content to stay that way, scraping a

poor living from a hard land at the end of the sea road. Beyond them, ashore, stretched the desert. New Zealand hid its beautiful wooded mountains and fertile plains beyond the trade wind belt well to the east of Australia, reachable only to those who approached it from the other side. Sailing canoes and praus could go to windward very well in quiet water and make good time in reasonably good weather.

But the Pacific was something else. The very name was measure of Europeans' ignorance of it. Pacific? The 'peaceful' ocean? Within the trade-wind zone – between approximately the Tropic of Cancer and the Tropic of Capricorn – at certain times of the year, perhaps, though even here there are seasons

of typhoons, cyclones and hurricanes. These are all names for the same sort of severe circular storm in which a sailing-ship of any sort was hard put to it to survive: a cyclone in the Coral Sea is the same as a typhoon off Guam or the Carolines, or a hurricane among the Low Islands. Beyond the Trade Wind belts are the variables of south and north Pacific, where anything can happen and the weather is far from settled. Beyond these areas again are the wild west winds, those fierce winds which, often at gale strength, roar unimpeded round the world in high southern latitudes culminating in the awesome storms of Cape Horn, and from latitude 35° North northwards, lash the North Pacific into a savage temper.

Only in the far south of the Pacific Ocean does the sea have an unimpeded 'fetch' right round the world – that is, the whole globe to roar round without any mountains in the way or land at all, south of Africa, Australasia, South America. Tierra del Fuego and Cape Horn dip well into this fearsome area like a divine breakwater to save the South Atlantic from the worst of the South Pacific's fury. On the Pacific side this causes the maddest reaction from the gale-driven sea, which flings itself upon the coast in a deluge of driving spume and stinging spray, tormented and demoniac, as if trying to wear down the land and make more room for its own savage passing. In winter, the days are short: in summer, the fresh-calved 'bergs from the Antarctic drift to the north, and they have not far to go before they lie in the paths of ships. Fog comes with the ice. In the north, Arctic ice comes through the Bering Straits and chokes the inlets of Alaska. Between Siberia and Alaska lie some of the coldest, stormiest seas even in the misnamed Pacific. The system of currents – equatorial, coastal, far south and far north – is of value to ships when understood and used, not fought. But Asian seamen knew only those waters that they sailed. These could be sufficiently difficult. It was left to the restless Europeans to wrest the Pacific's secrets. First of all they had to develop seaworthy ships, able to keep the open sea for periods previously undreamt of – months and months, even years. They had also to find officers to handle such ships, seamen to man them, and provisions to sustain these men. All this took centuries.

In both the North and the South Pacific temperate zones, speaking broadly, the wind and current systems go round and round and sailing-ships could go round with them, just as in the North Atlantic – sailing eastward in the temperate zones, west in the tropics. There were storms and – even worse in the days before engines – there were zones of calms. But long ocean voyages were feasible. It was the Europeans, led by Portugal, who found out how to make them. *All* sea-borne discovery belongs to the sailing-ship era, and by far the greater part was done before seamen knew how to keep accurate record of where they were or how far they had sailed. There was no reliable way to establish longitude. Among other obvious difficulties, this made it impossible to measure day's runs or to establish a ship's real position at sea. Drift was guesswork; knowledge of currents was gained only very slowly. There was no reliable way to measure ship's speed, either. It was left to judgement, aided by rough methods like the Dutchman's log, which simply meant throwing a chip of wood from the fo'c's'l-head and trying to measure the precise time it took to drift past the length of the ship, by chanting set phrases at set speed until it was abreast of the poop rail – a method requiring considerable optimism, though it was of some use later when good time-pieces were available.

It is no wonder that sailing-ships on long voyages were often a thousand miles and more out in their reckonings. The remarkable thing is that cockle-shells like the ships of Columbus, Dias, Cabral, da Gama, and Magellan survived their epic voyages at all – especially the handful which did battle with the Pacific in the region of the Straits of Magellan and Cape Horn.

Of the men who led expeditions in those waters the first was Ferdinand Magellan, a lesser Portuguese nobleman sailing in the service of Spain. Magellan was fortunate to get through the tortuous and squally straits named after him, and even more fortunate to find the Humboldt current comparatively close outside to help him sail to the north. His achievement was a great one and his voyage the greatest, perhaps, ever made by man.

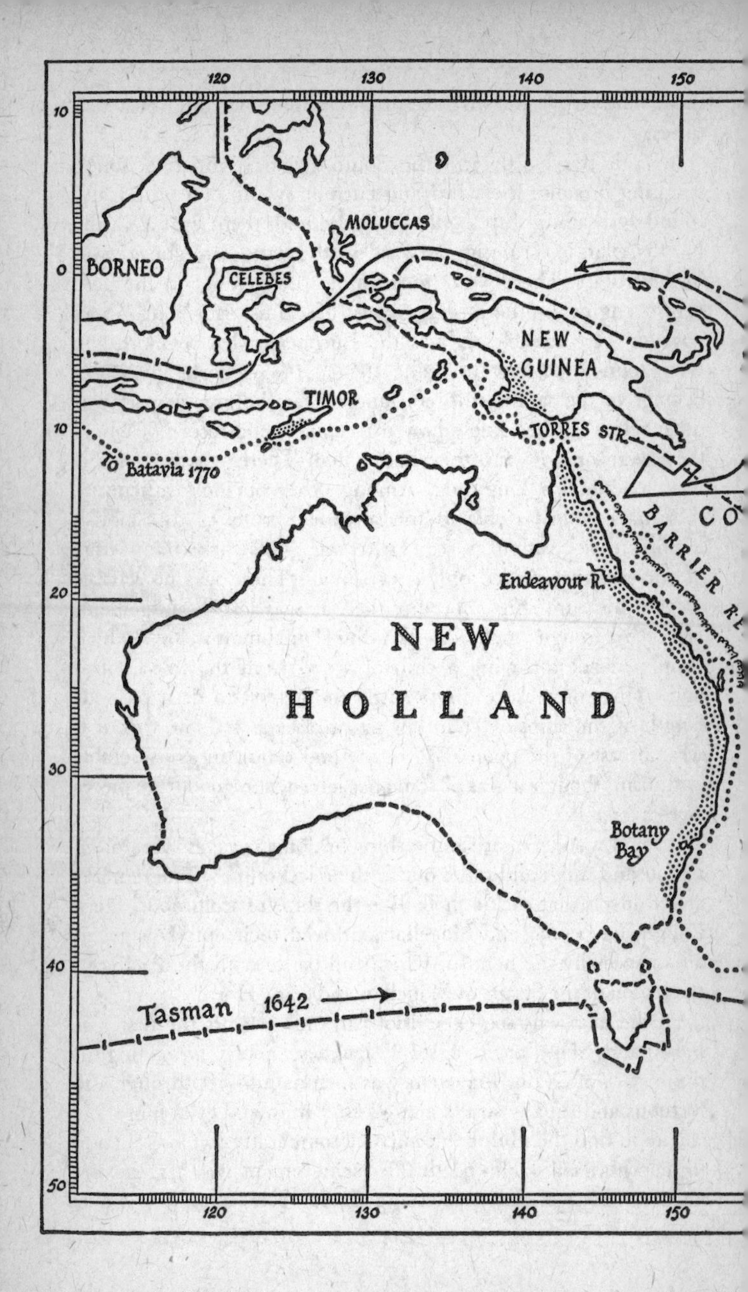
120
130
140
150
10
0
10
20
30
40
50
BORNEO
CELEBES
MOLUCCAS
NEW
GUINEA
TIMOR
TORRES STR.
To Batavia 1770
BARRIER
Endeavour R.
NEW
HOLLAND
Botany
Bay
Tasman 1642

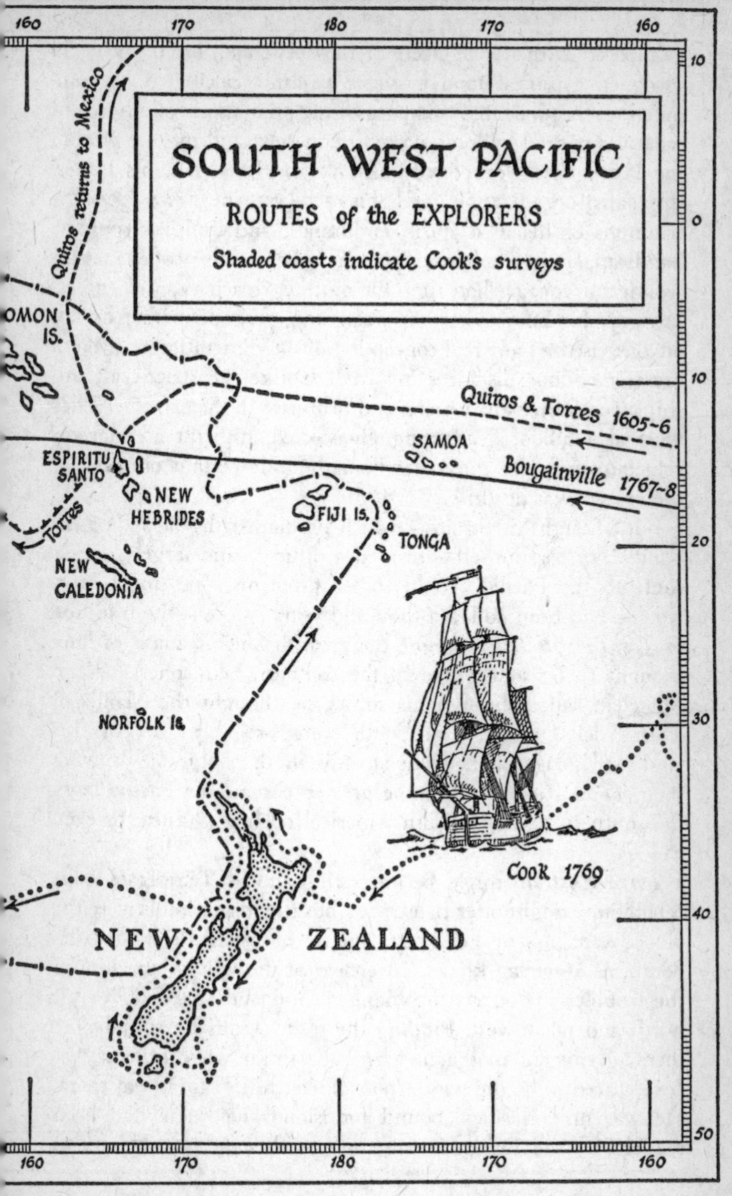
SOUTH WEST PACIFIC
ROUTES of the EXPLORERS
Shaded coasts indicate Cook's surveys
Quiros returns to Mexico
OMON IS.
ESPIRITU SANTO
NEW HEBRIDES
Torres
NEW CALEDONIA
SAMOA
FIJI IS.
TONGA
Quiros & Torres 1605-6
Bougainville 1767-8
NORFOLK IS.
Cook 1769
NEW ZEALAND
160
170
180
170
160
10
0
10
20
30
40
50

C.C.S.S. - 3

Compared with the pioneers of the spage age, briefly flung in cocooned capsules upon passages carefully calculated by computer and brought into existence by the joint efforts of a horde of scientists backed by long-massed know-how and infinite wealth, the Portuguese worked virtually alone. His vision, his leadership, his organization and his achievement were personal triumphs of his own spirit, endurance, and limitless courage, handicapped by numerous persons who, in the course of his wonderful voyages had time for mutiny, treachery, and endless intrigue. He knew where he was going, yes, but neither he nor anyone else had any real conception of the difficulties, or even of the tremendous distances involved. Unlike the space-craft, his ships were unwieldy, unsafe, and primitive in the extreme. They were also full of headstrong, ill-assorted, difficult and largely reluctant men, for most of whom the only certain outcome of the voyage was death.

But in truth his Straits – which he named the Straits of All Saints, not for himself – formed a difficult and largely useless route to the Pacific. To his other problems, one unnecessary trouble had been added a thousand years before – the tradition of *Terra Australis Incognita*, the great 'balancing mass' of land thought to be somewhere in the southern hemisphere. When Magellan sailed through his straits, he thought the islands of Tierra del Fuego to the south were probably part of this land, for geographers were already in the habit of drawing their *Terra Australis* over the greater part of the Pacific from the southern point of South America to New Guinea, or even to Java.

Terra Australis might be interesting and its Temperate Zone inhabitants might offer rich trade: but the spice islands were the objective of his voyage. They were in the North Pacific not the South, as Magellan knew. Once clear of the windy labyrinth of the troublesome straits, he thanked God and sailed off to the north and north-west. Finding the trade winds he sped before them, giving all that area where the unknown continent was conjectured to be the widest possible berth, in case it was there. He was on a passage bound for islands he knew had been reached by the Indian Ocean route – Ternate and Tidore, in the sea between Mindinao and the Celebes, and trans-Pacific pioneering, for the moment, was incidental.

Indeed, had he any means of knowing the appalling immensity of the distance he had to sail, it is perhaps doubtful if Magellan would have tried the voyage at all, or have found crews to follow him. He and his men ate everything in their ships they could, including the leather chafing mats from their rigging. Though the strips of leather were towed overboard for four days to soften them and then 'broiled' on embers for twenty-four hours, there was no sustenance in them. For ninety-eight days the decimated, scurvy-ridden little ships staggered across the unending, scornful sea. No storms blew. No rain fell. Oil and wine were at an end. The price of a thin rat was half a ducat. What little barrelled water remained was a slimy, nauseating stench which they swilled in their dry, gum-swollen mouths, with a firm grip on their noses. For unending misery and hardship that pioneering trans-Pacific voyage was the worst ever made (though a few others were later to follow something of the pattern – the Portuguese visionary Quiros from Peru, the Englishman Carteret from the Magellan Straits to the East Indies). Before Magellan was across he was 2,200 miles out in his reckoning; more than half his men were dead. In the end only one ship and thirty-one men survived to reach Spain.

Magellan's work was done and well done, though he died in the Philippine Islands in unnecessary battle for a treacherous native prince. He was already in fact the first circumnavigator when he reached the Philippines. The Portuguese reached the waters of the Western Pacific when they passed through the Straits of Malacca in 1511: Magellan, sailing out by the Good Hope route with the Viceroy Francisco d'Almeida in 1505, had distinguished himself both in battle and on exploratory voyages in those waters, long before entering the service of Spain. His pioneering trans-Pacific voyage was remarkable in many ways – to a sailor, perhaps above all for the danger-free route he took from his straits to the Ladrone Islands, for though a thousand coral islands and a million coral reefs were in the way he sighted none of them. He saw nothing but two small and sterile islands in the tropics somewhere, neither of which can now be identified with certainty.

Was this, one wonders, quite by chance? Or did he have some pre-knowledge of the way, if not the distance? One thing we can never know with certainty of these great pioneers is just what they themselves knew, *before* making their voyages: the sum total of their knowledge is their secret, and likely to remain so. Magellan himself had been sent by another great Portuguese, Affonso d'Albuquerque, on a groping voyage of Eastern discovery with that shadowy figure Antonio d'Abreu, or de Brew. It is considered probable that this expedition reached at least New Guinea. There were so many and so varied rumours and hypotheses about *Terra Australis Incognita* and Marco Polo's fabulous land of Locach or Beach (described variously as 'south of Java' or in Cambodia), islands of gold and so forth, that something had to be done about looking for them. It was early obvious that some land, *something*, must be somewhere in the South Pacific, and those fantastic empire-builders the Portuguese would have been indeed foolish not to look. If this *Terra Australis* – a huge land-mass largely in the temperate zone – really was inhabited by the same sort of people as in the north temperate zone, it behoved European pioneers to seek knowledge of at least the possibilities of attack by them. And they might offer opportunity for trade.

The reality of the reef-littered Coral Sea and the Great Barrier Reef of Australia put an end to dreams of a rich land waiting the happy day when Europeans arrived to exploit it. Any who reached that far – and there is an odd persistence in the belief, and at least presumptive evidence that some Portuguese did – found the South-East trade blowing hard in their faces, the sea a maze of reefs, cays, great depths and sudden, keel-clutching shallows, and such natives as they chanced to meet the most backward primitives on earth, grubbing a miserable existence, scarce able to keep their eyes open for a plague of persistent and most horrible small flies.

At least there was no prospect of attack from such a quarter. There might have been some prospect, however, of Spanish or other expeditions, bound from the Straits of Magellan to dispute the Portuguese trade in the East, to develop north-east Australia as a convenient base to refresh ships and crews, and to launch attacks from there. Suspicions of this sort would be more than sufficient to account for so thorough a shield of Portuguese silence that it has not been broken yet. There were, after all, so few Portuguese; the Pacific and the East were so enormous; and the Spice Islands were so rich.

It is an odd fact that by 1530, there were charts produced in France, using filched Portuguese data, showing not an imaginary *Terra Australis* but something very like the real Australia where Australia actually is, marking the Coral Sea as dangerous, with indications of Queensland's Great Barrier Reef and the coast of New South Wales.

As for the sailing-ship route from the Atlantic to the Pacific through the Straits of Magellan, this was of little practical value. For the Portuguese it would have been far better to find a way to the Indies round North America, not South. A straits or passage through Canada from Hudson's Bay would at least have been open in the summers, and was so much shorter.

But there was no such convenient opening.

Magellan's principal contribution in his own times to the new ocean was to demonstrate its immensity and its essential uselessness. It was too difficult a route for Spanish galleons trading from Central America and Peru to Spain – better to use local-built coastal craft, kept in the Pacific, to carry this trade to the Isthmus of Panama. There it went by mule train to the Atlantic side for onward shipment. Magellan discovered the Philippines, too: in time a trade was developed from there to Spanish ports in Mexico. In 1565, Andres de Urdaneta found the way to sail eastwards to Mexico by first making northwards beyond the trade wind belt, then eastwards with the predominantly westerly variables. It was simple enough, though a lengthy process, to sail slowly before the north-east trades the other way. Captains of galleons were reluctant to take those lumbering great ships – so beautiful but so awkward – out of the trade wind zone, but, until one did, the eastbound passage of the Pacific was impossible. Urdaneta's contribution was largely commercial, but it was a considerable one.

In the South Pacific, shortly afterwards, another *hidalgo* – son of a somebody, scion of the nobility – set out from Peru to test an Inca legend about 'islands of gold' 600 leagues to the west-'ard in the Great South Sea. This was Mendana, son of the Viceroy. With him went the able Sarmiento de Gamboa. They sailed not 600 leagues but some 2,000 and came eventually on a group of high, humid islands which they named Guadalcanal, Malaita, San Cristoval, and Ysabel. These became known as the Solomon Islands – at least to waterfront rumour in Mexico and Peru. The names stuck: but the islands did not. It was another thirty years before, in 1595, Mendana tried to find them again, this time to colonize them. He failed, but his chief pilot was Pedro Fernandez de Quiros – another Portuguese, like Magellan. Quiros was a visionary with some ambition to become the Columbus of the Pacific. His theory was that the Solomons were some of the fringing islands of the long-lost *Terra Australis*, just as the West Indies proved to be for the Americas, and he felt a visionary urge to discover them and *Terra Australis* with them.

On the 1595 voyage Mendana and Quiros sailed through the Marquesas and on to Santa Cruz, but the Solomons eluded them. Still with very hazy ideas of just what an enormous distance they had sailed, Mendana died. The surviving riff-raff, male and female, showed no disposition to colonize anything. Quiros was forced to abandon the enterprise. He sailed on round the north of New Guinea to Manila in the Philippines, whence the survivors picked up Urdaneta's route to Mexico and thence down the American coast to Peru.

Now more than ever determined to find the elusive *Terra Australis*, Quiros petitioned urgently for permission to make yet another voyage. In 1605, he was allowed to go, for that – quite baseless – legend about the Isles of Solomon had spread. The inquisitive English, whose marauding Drake of Devon had made his great circumnavigation thirty years earlier, were reported to be interested in it.

So off sailed Quiros again, on yet another interminable passage westwards in the South Sea, with yet another of those frail little fleets manned largely by waterfront hoodlums and other riff-raff the discoverers were so often fobbed off with, on the locally applied principle that their voyages were probably of no use anyway and therefore the less at stake the better.

This time Quiros, with his fellow Portuguese Luis Vaz de Torres, being in a slightly different latitude, came in time – a long time – to the island group we know as the New Hebrides. Hailing these malarial, hot and mountainous islands as *Terra Australis* itself, the ecstatic Quiros named them Austrialia del Espiritu Santo, chose a site for his city of New Jerusalem, set up an Order of Knights of the Holy Ghost, and quite lost his sea-going senses in an orgy of religious pageants and ceremonies. Without finding out even that this Austrialia etc. really was just another group of South Sea islands, the deluded navigator soon afterwards found himself on the old Urdaneta track sailing back towards America, whether voluntarily or not we do not know. It is thought not. No one ever entrusted him with ships or expedition again. It was a pity, for Quiros the disinterested visionary *had* made great voyages in the course of pure discovery. If there were any *Terra Australis*, he was a fit man to find it.

Torres parted from his too-ardent leader and sailed 400 miles to the south-west first, looking for *Terra Australis*. If he had sailed 1,200 miles he would have reached Australia, perhaps. He would also have got among the coral of the Great Barrier Reef. But he turned away like a man with some idea that the coral was there. He turned north-west and west-north-west, and proceeded on a very curious course indeed. For Portuguese Luis Vaz de Torres, a Peruvian Pacific pilot who had never been in those waters before, calmly sailed along the southern coasts of New Guinea, though his little ship was foul, poorly manned, and unfit to beat back to windward if she must. Small square-riggers – or large – with rope rigging and light wooden yards could not stand up to the endless slogging strains of constant tacking, expecially on long open-sea voyages. Torres *had* to be certain of that passage, which now bears his name.

Who told him? Who had been there? Most probably, one of those shadowy figures of the then recent past, a Portuguese like d'Abreu or perhaps Heredia. What else did they know? Enough to have 'leaked' the data for the Rotz map? When Torres sailed through his straits he was very close to the northern point of Australia, within sight of it. He doesn't mention it. Did he know about that, too? If he did, he didn't bother to go there, and that was understandable, too.

Before this, another group of intrepid and able sea discoverers had followed up the Portuguese sailing routes by way of the Cape of Good Hope and the Indian Ocean. By 1595 the Dutch were sailing their own ships towards India and Indonesia, regardless of the Papal decision dividing the Asian and Pacific worlds between Spain and Portugal. A few years later the Dutch East India Company was formed, and soon prospered greatly. It was as obvious to the Dutch as to the Portuguese that, as the Pacific route to the Spice Islands was known, interlopers could come that way and cut in on their trade. Their East India Company had a monopoly by its Dutch-granted charter of the established sea routes round Good Hope and through the Straits of Magellan: was that enough? What about this *Terra Australis* as a wayside stop, or worse, base for English and other marauders? Their English agents had reported that Drake was interested in the idea but found enough without. That country could produce more Drakes. The Dutch monopoly was not strengthened when their countrymen, Isaac le Maire and

Willem Corneliszen Schouten, dissatisfied with the difficult Straits of Magellan, had the effrontery to set up an 'Australian Company', and to sail the usual couple of cockle-shells from Holland to the Pacific by a new way. This was past the Horn, the southern extremity of all the Americas. It was common talk among seamen that Drake, driven back that way after getting through the Straits of Magellan, had spoken of a 'main ocean' with tumultuous winds and seas, which indicated the absence of any considerable land.

Passing Cape Horn (which they named 'Hoorn' for the town of Hoorn in Holland) Le Maire and Schouten found themselves in Drake's great sea where 'sea-mews larger than swans, with wings stretching a fathom across, flew screaming round the ship'. These alarmed the more superstitious (who were not exaggerating: the albatross has a wing-span often of more than two fathoms) who regarded them as the spirits of the South Sea and wanted to turn their ships away from the westerly gales and go back to Holland. Later seamen beating down there were to regard those 'sea-mews' as reincarnated old sailors, and view them with benevolence. Others, without superstition, were to eat them.

In due course, the surviving Le Maire–Schouten ship – the *Hoorn*, a yacht, was lost – sailed through the Pacific not far from Magellan's route, crossing the tracks of Mendana, Quiros, and Torres. Le Maire wanted to stay in the South Pacific to find

Terra Australis, but Schouten and the ship's council forced him to leave the area and make for the East Indies, where Schouten had already been. He knew there was rich trade there, readily convertible into gold. So they passed north of Espiritu Santo and the Solomons, by-passed the real Australia, reached Java north-about past New Guinea, and were promptly flung into jail as lying interlopers whose claim of a 'new route' was a fraud.

They *had* discovered Cape Horn and, with it, a better way from the Atlantic to the Pacific – at least for strong sailing-ships. But the Dutch East India Company need not have worried, for the new route, as far as reaching the Pacific was concerned, was more deterrent than aid, with or without 'sea-mews'.

The Le Maire–Schouten voyage was in 1615. Almost a decade earlier, while Torres, profiting by knowledge unknown to the Dutch, was sailing westwards to the south of New Guinea making for his straits, another Hollander was probing painfully eastwards towards – he hoped – the other side of Torres Strait, or any other way through to the Pacific that he might find. This was Willem Janszoon, who sailed from Djakarta – then Batavia – in a handy little fore-and-after named the *Duyfhen*, in 1605, with explicit instructions to explore the *island* of New Guinea. The Dutch Governor reasoned that if Indonesia was returning average annual profits to the Company of almost forty per cent, the big province of New Guinea being in the same general area could be expected to offer profitable trade, too. But Janszoon got the *Duyfhen* embayed in the Gulf of Carpentaria and did not find the western end of Torres Straits at all.

He *had* seen Australia. To him the place was a desert inhabited by 'wild cruel black savages' who, caring for interlopers no more than the Dutch, murdered some of his crew. Another Hollander, Jan Carstenz, sent on the same quest in 1623, did no better. His speciality was trying to kidnap aborigines in order to teach them Malay and so learn more of *Terra Australis* from them. This was optimism indeed. After a few attempts, Carstenz was surprised to find (says his log) that the 'blacks received us as enemies everywhere'.

So he sailed back to Batavia, too.

In the meantime Dutch shipmasters had found the real

sailing-ship route to the East – the west winds passage which later sailors were to know as the 'Roaring Forties'. Observing that useful westerly winds prevailed south of the Cape of Good Hope, an enterprising Hollander kept his Java-bound ship down there in those strong, favouring winds, and let her blow along eastwards to the longitude – more or less – of Java. That reached, he turned north to find in a few days the Indian Ocean's south-east trade winds. These blew him right to Sunda Straits. The previous route was to sail diagonally over the Indian Ocean, or to go north along the East African coast and then strike across. These were slow routes, requiring strongly-held bases in East Africa and Arabia while making ships vulnerable to marauding Moslems and others striking out from those waters.

The Roaring Forties way was a considerable improvement, but it demanded strong ships and good seamen. These the Hollanders, raised in the tough waters of north-west Europe, could supply. But it demanded also a certain precision of navigation which was then beyond them. Those westerly gales often picked up a ship, flung her along in the wild sea where it was as much as she could do to pick up her skirts and run – and run and run as ships had never run before. Who knew how far? Four thousand miles? There was a lot of the East Indies, strung obligingly along well over a thousand miles of longitude. But who could measure that? Striking north after making four or four-and-a-half thousand miles from the Cape, a ship *should* reach some recognizable spot among the Indies, and then coast along to the Straits of Sunda or of Bali.

But the real *Terra Australis* was in the way if you ran too far, and Dutch seamen soon found it. The west coast of Australia too often wrecked their over-running ships. They called the place New Holland and found it deplorable. The more they saw of it the more were they confirmed in this pessimistic view.

Before long, the western and much of the north-western, northern, and southern coasts of Australia were on Dutch maps, at any rate in broad outline. What about the east? Where was that? Joined up somewhere with New Guinea to form *Terra Australis*, reaching down to the Antarctic, or half-across the

Pacific? If Carstenz and Janszoon could not solve the puzzle by passages along the north of New Holland, then perhaps bolder mariners staying in the west winds of the far south until they *did* reach land should find out something.

So off went the courageous Abel Janszoon Tasman with Frans Visscher, a 'pilot of renown', sent with two ships by Governor Antony van Diemen of Batavia. Here was a voyage of discovery well worked out, with good men and satisfactory ships. To miss nothing on this long Roaring Forties run, Tasman was ordered to sail to Mauritius first – the island was then a Dutch port-of-call – in order to be well to windward, then get down to 47° South or so, head east and run. He could run as far as the presumed longitude of the lost Solomon Islands and 800 miles more, if no land intervened. He was to check on these Spanish discoveries, to see whether they were connected with *Terra Australis* or any other great land. In short, Abel J. Tasman was given a very tall order indeed, which could have led to the Dutch solution of most of the mysteries of the Pacific.

Off to Mauritius he sailed, bounding across the Indian Ocean with the gentle trade winds on or abaft his port beam. Flying-fish skimmed away before the homely bows of the little *Heemskirk* and *Zeehan* and the mariners basked in the benevolent sun, for this was one of the most pleasant sailing runs seamen could ever know. These Indian Ocean trades were more reliable than those of the western Pacific. Day after day the ships ran on, rolling gently, their sprits'ls dipping softly in the roll of foam running in broad, admiring lines from their bows. For weeks the wind was so constant it might have been designed to seamen's specification and delivered by their order. It was indeed the Lord's gift to the intrepid sailing-ship seamen. They deserved it.

Soon the *Heemskirk* and *Zeehan*, the flying-fish weather over for the time being, were leaping along in the great seas down towards 50° South latitude. Sometimes they seemed to try to jump right out of the tormented sea, leaping from some breaking, rolling, noisy crest only to slide into the trough while the sea rushed on and the quartermasters wrestled with the whip-staffs, trying to keep the ships on something like their course, until their arms ached. Those small, round-bellied wooden ships were buoyant, rising and prancing in the sea instead of floundering in it as deep-laden steel sailers were to do later. Sometimes great seas smacked their fat windowed sterns, but the seamen had built wooden shutters over the windows and little water got in.

It was uncomfortable, but it was safe – well, as long as neither unknown land nor drifting icebergs, up from the nearby Antarctic, got in the way. It was spring and early summer when Tasman raced through the Roaring Forties, the season when giant icebergs and fields of pack, cast loose from the frozen continent of Antarctica, drift towards the north. The pack-ice does not persist for long, disintegrating in the sea, but some of those Antarctic 'bergs are more like ice-islands, twenty and even fifty miles long. Tasman eased his ships' rush by night, but he had to keep some way on if only with the lee corner of a sprits'l set, or they would wallow their masts out in the trough of the sea. Trebled lookouts kept straining watch ahead and listened with straining ears for the sound of breaking seas – breaking not in their own tumbling crests, but on land, or ice. Even a small sailing-ship makes a lot of noise with the wind's roar and the sea's shout in a gale or fresh wind. She is a very low platform for lookouts in a big sea.

Tasman's mariners, for long used to tropic seas, found the Southern Ocean trying: so he hauled to the nor'rard and ran on the 44th parallel. After forty-six days, they came to Tasmania. Tasman – or more probably Visscher – worked out his longitude by lunars and dead-reckoning as 140° 46′ east of Greenwich which was within thirty miles of the actual position of the west coast of Tasmania. He took possession of the 'said land as our lawful property' by right of discovery, made as good a running survey as he could, saw enough to convince him that the mountainous land was as Nature had formed it, offering neither teeming villages nor trade, and hurried on. Over the Tasman Sea the *Heemskirk* and the *Zeehan* ploughed their way, fighting with the sudden shifts of stubborn squalls that stab at those waters, driven from their chosen latitude of 41° towards the north by storms which Australian seamen later knew and dreaded as 'southerly busters'. That Tasman Sea is a wild place, a segment of the South Pacific far removed from the tranquil idylls of the South Sea. Tasman, Visscher and their men, pioneering the hard way, were more heroic than they knew or their contemporaries ever gave them credit for.

Sailing that way, they came inevitably upon New Zealand, a mountainous land of wild grandeur rising abruptly from the sea, and they were the first Europeans to set eyes on it. Tasman approached the strange land warily, aware that it is purely a landsman's view that land is safe after sea. A good harbour may be so, but first he had to find one. Dangerous lee shores are far more abundant than harbours, good, bad or indifferent, and even a 'safe' harbour can be much impeded and imperilled by shoals, sunken rocks, and the like. He shortened sail prudently, got the cables ready on the anchors, and his primitive leadlines prepared for taking soundings.

The sort of ship which Tasman had – and not just Tasman: *all* the sailing-ship discoverers – was extremely vulnerable near the land. Caught by unseen sets and unpredictable tides, she could be swept ashore if she came too close in. She could become embayed – unable to beat out of a shallow bay because she could not weather rocks or headlands on either tack – and if her ground tackle did not hold or there was not a favourable shift of wind, she was done for. Anchors were primitive contrivances then and cables were of hempen rope, designed to be chewed away on rocks or coral: the means of working anchor, by hand-operated windlass and capstan, were very primitive, too. A discoverer's work was far from done when he made safe landfall: indeed, it was only just beginning. Seamen used to deepsea voyages often dreaded the land, with good cause.

And so Tasman kept his ships as a convenient distance from the strange coast – near enough to see townships and villages, if any; far enough not to risk being embayed or otherwise put in grave danger. He saw neither towns nor villages, nor much sign that this land was any different from primitive Tasmania. A few Maoris, coming out in a double canoe, failed to comprehend anything bawled to them by the Dutch interpreters who used a pirated copy of a dictionary of sorts made by Mendana's men in the Solomons over half a century earlier, and some phrases picked up by other Dutchmen on the coasts of New Guinea. The Maoris were Polynesian, and neither the Hollanders nor any other Europeans had any knowledge then of that melodious language.

So there was no point of contact. The wary Maoris refused to be tempted aboard the *Heemskirk* by a show of cloth and trinkets. Not long afterwards, a further meeting with Maoris led to sudden death – the first such meeting. It is fair to say at once that it was Dutchmen who died and the Maoris, apparently, who were treacherous. Tasman simply wanted to find out what they knew of the new land. The Maoris didn't want them to find out anything, taking a poor view of strangers arriving by sea. They still had a tradition of having come themselves in the same way before wiping out the original inhabitants. At any rate Tasman anchored his ships close in-shore: one of his boats was run down, apparently deliberately, by a large canoe, and four Dutch seamen in it were clubbed to death.

Tasman called this place Murderers' Bay, recovered his boat, hove up and departed. As he sailed, a dozen canoes manned by warriors put off from the beach towards the ships but a few Dutch shots put a stop to them. It was a pity. The Maoris, though sometimes given to eating strangers who came among them, were an intelligent race who knew something about voyages and about the Pacific, too.

Part of Tasman's instructions required him, if not stopped by *Terra Australis*, to find a clear passage towards Chile and Peru. His ships were actually in Cook Straits, through which there is such a passage. But this Strait is a baffling place for square-rigged ships. After sailing before westerlies for a while, these turned to headwinds just where the coast of the North Island trends southwards, seeming to block the way. An easterly wind later gave Tasman the chance to get out again from this 'baffling bay', and he got out. His was that same old problem of square-rigged ships in danger of becoming embayed. If there were no passage, and he allowed his ships to become jammed in the lee-ward end of the bay, what then? Shipwreck, most likely: *and* among murderers. He *might* beat out, of course, but sailing-ships of the seventeenth century (and the eighteenth) were crude vessels which could not keep the seas indefinitely or beat very successfully against persistent strong winds. Cook Strait is in the zone of the Roaring Forties.

Tasman did the seamanlike thing. He had other objectives set by the governor at Batavia. What about those Solomon Islands, and Quiros's Espiritu Santo?

He named the new land Staten Landt and its northern extremity Cape Maria van Diemen (for the governor's wife), expressed the view that his discovery was the west coast of *Terra Australis* (though Pilot Visscher was still doubtful about that big bay), and sailed on towards the north. His all-embracing instructions included also a directive to find and examine the eastern coast of New Holland. Poor Tasman! No man could have done all that Governor van Diemen set him to do – find *Terra Australis*, the extent of New Holland, a passage to Spanish South America, the gold of the Solomon Islands and any other gold or trade that offered.

Set to find so much, he missed the lot. But he made great discoveries – Tasmania (which he called Van Diemen's Land), New Zealand, Tonga and the Fijis on his way north. He made no attempt to seek New Holland from the Pacific side, though he had pinned down roughly where its east coast should be sought – his voyage was in fact a circumnavigation of Australia. He returned to Batavia by the more or less beaten track beyond the New Hebrides, Santa Cruz, the Solomons, and round the north of New Guinea.

Tasman and Visscher, as discoverers, had indeed done well. But Governor van Diemen was not pleased. The cost had been considerable; the profit, nothing.

Tasman was sent again in 1644, this time on the old track north of Australia, with instructions to pass through Torres Straits and then sail down the Australian east coast as far as Tasmania. But, like his countrymen Janszoon and Carstenz, he did not find the entrance to Torres Straits, nor any other channel into the Coral Sea.

If he had, what then? For the south-east trade wind would be right in his face, blowing at him over the nastiest maze of coral reefs in all the world – thousands of square miles of them. He would be embayed indeed. Perhaps Tasman, Carstenz and Janszoon – or the pilots assigned to work with them – were not really all that keen to find a coral-free passage over the sunlit but so treacherous sea between northern Australia and New Guinea. To find such a channel and pass through would be not an end but a ghastly rebirth of all their troubles. It was their lives which were at stake.

I once, in the 1930s, sailed a full-rigged ship that way, the iron *Joseph Conrad*, from Samarai in Papua to Lord Howe Island, bound towards Tahiti. Samarai is several hundred miles beyond Torres Straits and outside the Great Barrier Reef. All I had to defeat was the Coral Sea. I had as good charts as I could buy (though these, even then, were far from perfect), a chronometer and a modern sextant, an iron ship with wire standing rigging, a good crew. Yet it took me six weeks to make a few hundred miles to wind'ard. Those reefs were not lit: the positions of many were vague. They were awash, hard to see by day, impossible by night. I tacked ship when I saw a reef by day or heard one by night. The sound of the breakers snarling upon that ghastly sword-sharp and pike-toothed coral was unmistakable and alarming: my crew and I heaved great sighs of relief when we were gone from there. I never returned.

Poor Tasman, indeed. He had made a great voyage: after him, no other Hollander sailed into those waters. Holland's seamen had enough to do handling the large ships in the Indian and Indonesian trades, pressing on (after the Portuguese) to China and Japan, making voyages of some risk and great profit for the best part of the next three hundred years.

Terra Australis Incognita, the shape of New Holland, the rediscovery of the Solomon Islands, the investigation of Tasmania and New Zealand – these things could wait. It was scarcely likely that any of them would rival the fabulous trade of the Indies. As for threats and fears of invasions, the years passed and these things never came.

Maybe the vast Pacific was, after all, the scar upon earth left by the loss of the moon – enormous, barren, and, except as a sea track for the hardy, probably useless. If the moon had ejected itself from there, it was reasonable to think that it had taken all the land with it.

CHAPTER FOUR

The Literary Discoverers

NEITHER Portugal nor Holland was interested in Asia, or the Pacific, for the establishment of colonies. They had no surplus population for such expensive enterprises, nor need of them. They had indeed no surplus populations at all. Plague, pestilence, poverty and wars took care of that. What ranked high in North European objectives was trade, monopolistic if possible. The Portuguese, spread thinly over the vast area of Afro-Asia, clinging to sea power here and there round the fringe, could not maintain monopoly.

Since they were Atlantic summer and winter seamen and not Mediterranean summer plodders, their ships were better – more seaworthy and more stable gun-platforms – than the Arabs, Indians, and Malays. But the northern European Dutch, English, and French had, in time, developed even better ships. So in due course, disregarding the Papal division of the world between Spain and Portugal – a somewhat ambiguous affair at best – one after another the Dutch, English and French tried to move in, succeeded, and spread. The problems of the Pacific, now the sailing routes into it and across it were known, could wait. A few odd fellows might wander in the great South Sea for their own reasons – privateers, buccaneers, interloping would-be merchants and the like. The principal products from these, apart from the proceeds of raids on the odd treasure galleon, were books – travel books.

The first travel-book writer was an unusual Englishman named William Dampier, born in 1652, the son of a tenant farmer near Yeovil in Somerset. Young Dampier thought little of farming and went off to sea as a professional seaman, first in the Newfoundland trade – salt out, dried cod back – and then on an East Indies voyage. He cared little, very understandably, for either. The Newfoundland voyages were too cold and hard, the East Indies too hot and too frequently fatal, for more than half the crews died of fevers ashore or scurvy afloat. Dampier was a bright youth though restless, and very observant. He was something of a mathematician too and, as such, learned what he could of navigation. He also was a forerunner of Captain Maury, and learned all he could about the behaviour of ocean

Becalmed

winds and the making of ocean voyages. A spell at the age of twenty as seaman in the Royal Navy was followed by an introduction to buccaneering in the West Indies, where he went first as an assistant manager of a plantation in Jamaica. Mathematical skill was probably of use in this business too, but Dampier soon found better outlets navigating for the more ambitious buccaneers on such jaunts as a march across the Isthmus of Panama followed by a piratical cruise in a Spanish ship seized on the Pacific side. His companions and he returned to Chesapeake Bay – a good place to hide at the time – to share out the loot.

Here, in 1683, a friend named Cook found a brig lying conveniently ready for sea and, for the moment, neglected: so back to the Pacific in the brig they went, to cruise off the west coast of South America. Cook died: one Davis took over – the same Davis who declared he had seen the coast of *Terra Australis* far out to the west'ard of Chile, in a part of the Pacific where more diligent or, probably, more sober navigators have been able to find only Easter Island. Splitting forces, Dampier crossed the Pacific in a little tub of a vessel prettily named *Cygnet*. He sailed by the beaten track to the north of New Guinea, touched at the Ladrones to refresh, cruised a while in the East Indies, and then called at New Holland. It would seem that Dampier was the first Englishman to see New Holland and the first seaman deliberately to seek the place. The others had been wrecked there, or sailed desperately away (if they could) as soon as it was in sight.

Dampier's men were not impressed, for the coastal area of West Australia round Broome and Port Headland seemed as barren and useless as the Dutch said it was, and the few aborigines they met were quite as hopeless. The crew of the *Cygnet*, bored with the tiresome tastes of the strange seaman from Somerset – he spent a great deal of time writing up journals and things but gathered no pieces of eight – threw him out of the ship at the first convenient spot, which was in the Nicobar Islands. Escaping from this predicament by canoe to Sumatra, Dampier became master-gunner of the English East India Company's fort at Bencoolen, for he was a many-sided and gifted young man. All this voyaging profited him little. When the chance came in 1691 for passage back to England in an East Indiaman and he went back after an absence of twelve years, the only 'investment' he brought was a tattooed Malay who was to be an exhibit at country fairs as a sort of 'wild man from Borneo'. But the Malay tired of this, picked up smallpox and died at St Giles's fair at Oxford. This was a blow. Dampier, ever resourceful, had one trick left. He got out his logs and wrote a book. Dedicated to the President of the Royal Society to establish its respectability just in case some of those buccaneering and other piratical exploits were remembered, and written by a very intelligent man with a nice turn of phrase, the book – *A*

New Voyage Round the World – was something quite fresh on the English market. It ran through four editions in two years and became the first travel-book 'best-seller'. Delighted with this, Dampier read further in those useful logs and produced two more books, which also did well. One of them was his *Discourse on Winds*. This was no travel-book but a valuable contribution on a little-known subject. Dampier had 'been around', had observed scientifically and preserved his observations. For a while he was a bit of a literary lion in London, for he was a personable chap with considerable charm.

The content of his books stirred English imaginations. This was the end of the seventeenth century, and two new worlds were calling thinking men in Europe – the adventurous world of new lands, new seas, even of a possible new continent far off in the great South Sea and, at home, a stirring new Europe vibrant with fresh ideas. The time had gone when the fruits of discovery were too jealously guarded. At least abstracts of accounts of most voyages were available to the learned. Tasman's tracks and discoveries were laid down on the floor of Amsterdam's Town Hall* when it was rebuilt in 1648. In England, when Dampier's books were published, the Royal Society had been founded for nearly forty years and Isaac Newton had been a professor at Cambridge for twenty. The minds of educated men were readier than they had ever been for new concepts, new ideas.

In these stirring times, My Lords of Admiralty had a brilliant idea. They would organize and send out an expedition to the ends of the earth, to sail Balboa's South Sea and unravel its secrets, to plant the British flag on *Terra Australis* which would be allowed to be *incognita* no longer. The leader of this one-ship expedition, the first exploratory voyage the Admiralty ever organized, was to be Dampier. He was not a naval officer, of course: but who else could be chosen?

It sounded wonderful. In fact, it could hardly have been worse, for buccaneer-scientist-writer-navigator Dampier, though

*Now the Royal Palace. The world map and the discoveries were still there when I called in September 1966: but the track – once laid in copper strips on marble – is gone.

excellent in all these fields, lacked the divine spark of leadership and seemed to have no power of command at all. His plans were fine and his intentions likewise, for he meant to beat his ship past the Horn and across the Pacific from there, not in the balmy trade winds but further to the south, in the latitude of New Zealand and Australia, in quest of *Terra Australis Incognita.* He needed two ships and three years. My Lords, taking no chances, perhaps, with the non-naval commander of their first naval exploring expedition, gave him one ship, the rotten old *Roebuck*, which was so ripe it was a miracle that she lasted one year. She was old, neglected, and decrepit, her planking hanging on its many coats of tar, much of her decking spongy to a barefoot tread, her stumpy masts and ill-fitting yards kept aloft by a combination of paint, varnish, a profusion of wooldings, and optimism.

Dampier tried to refuse her but he had already turned down a packet even worse, misnamed with grim humour the *Jolly Prize*. The *Roebuck* had been damaged at one stage of her long career when acting as a fire-ship, a service usually assigned to decrepit vessels of use for no other. It dawned on Dampier that someone in Admiralty was taking no chances and giving him none, either. However, there the *Roebuck* was. After the mockery of a brief refit where lengthy rebuilding would not have made the ship seaworthy, Dampier took her to sea. It was January, 1699, the depths of midwinter. Doubtless there was a useful easterly slant to get down-channel, perhaps as far as Finisterre – a good winter 'high' over Siberia or the Baltic spills off easterlies and north-easterlies, of great value to outward bound ships – or the *Roebuck* would never have got out of the English Channel. There was one thing (among many) she was not fit for, and that was to weather Cape Horn. One look at her and Dampier had given that idea up. Her voyage must be by the Good Hope way both out and back (if she ever got back) with as much time in the tropics as possible. Instead of approaching *Terra Australis* from the east, he would have to make towards that useless west coast of New Holland once again.

There were other, more serious handicaps, which soon became apparent. The crew suited the ship, for they were mostly incompetent and unanimously mutinous. The officers were little better. They all knew that their commander's sole naval experience had been a brief period as a seaman, and that he was an ex-buccaneer. They were not so sure about the 'ex'. A First Lieutenant named Fisher, bitter at serving under such a man, was openly critical alike of Dampier's seamanship and navigation. Whatever else he was, Dampier excelled in both these fields and any but a born fool or deliberate trouble-maker could see it. The leaky *Roebuck* rolled along, Fisher sneering, the crew alert for a chance of successful mutiny. Dampier took to sleeping on the deck of the Master's house right aft, one eye half-open and his pistols at the ready. Up there, he was hard to get at: but he could not tolerate the constant subversion of his authority. He put in at the South American port of Bahia and had Fisher flung in a Portuguese prison.

After this the crew were a little better. Dampier sailed again and, to take no chances, omitted the usual call at the Cape of Good Hope. Instead, he stood directly for the west coast of New Holland, which he knew where to find. Here the crew, slightly subdued by the west winds' run in their elderly sieve and alarmed at the sight of the desert New Holland, behaved themselves for fear of being marooned or – worse – wrecked there. By this stage it was obvious even to them that this odd fellow Dampier knew his business as a seaman.

It was equally obvious to poor Dampier that he had to give up his last hopes of making a worth-while voyage. He had meant to run along the south coast of New Holland which would have brought him to the straits between Australia and Tasmania, with the east coast of New South Wales just round the corner, and the Tasman Sea before him. Indeed he would have forestalled James Cook, then unborn.

Instead he had to turn north, to sail (if he could) round New Holland north-about – the hopeless way to reach its eastern coasts, as Carstenz, Tasman and others had found out. Off New Guinea's northern coasts he found a land which 'afforded a very pleasant and agreeable prospect'. This he named New Britain. Thankful to have got so far, he turned the rotten *Roebuck*'s fat prow homewards by the way she had come. She could no longer beat against the baffling trades blowing in her face. Having omitted the Cape, Dampier was short of food and stores, too.

The old waggon sank from plain inability to remain afloat, on the homeward passage, when planking round her leak came away in her shipwright's hands as he tried to repair her. The wood disintegrated and the sea rushed in with a joyful gurgling, pleased to be swallowing at last so long withheld a prey.

It was a beautiful morning within sight of an island – Ascension, in the South Atlantic, on the sailing route to the Line. Dampier and his men rowed ashore. A few weeks there and an English squadron picked them up (by chance) and took them home.

In England First Lieutenant Fisher, released from Bahia jail and more bitter than ever, had been raising all the hell he could for months. He was good at raising hell. Dampier was court-martialled, found guilty, fined the whole of his promised pay, and declared for ever a person unfit for command in the King's ships. Such unfitness might have been foreseen; he did not appoint himself, and he had earned that pay and his country's gratitude too. It was a mean court – stacked with four admirals and thirty-three captains, all career officers, none for him – and a mean sentence.

Well, they could keep their pay. Dampier's books still sold. He had good material for another. In any case if he were now unfit to command the King's ships, there were others where unduly truculent first lieutenants could be strung up at the yard-arm or made to walk the plank, and no courts martial held for them. Dampier was at sea again within a year, in command of a couple of privateers (with the King's commission) with conspicuous lack of success. This time he was concerned with the marooning at Juan Fernandez of that Crusoe prototype, Alexander Selkirk. Yet another privateering run – again to the Pacific, this time not trusted with command but, far more usefully, as navigator to one Woodes Rogers – led to fortune but not for Dampier: the expedition took 200,000 golden pounds from sundry treasure-galleons, and released Alex Selkirk as well. But the share-out was delayed by litigation and quarrelling for years. The English East India Company objected to the seizure of prizes within 'its' waters, and there were quarrels among the backers. Woodes Rogers and Dampier returned to England in 1711. It was Dampier's third circumnavigation. Three years later he was dead.

Though he obviously was an unfit person to command King's ships and was too scientifically minded to make a success of any others, Dampier was a great navigator who could have been a great discoverer, too. Something was missing in his make-up. Apart from New Britain, an island or two, and some useful knowledge of ocean passage-making, he left the Pacific as he found it. But his books and writings lived after him: they had a considerable effect.

So had several other books produced as accounts of voyages more or less accurate, or ruminations and discourses on the whereabouts, wealth, etc. of that alleged and most elusive southern continent, *Terra Australis Incognita.* Dampier succeeded in bringing the Pacific to men's minds, not only in England. Woodes Rogers produced a good book about his voyage. The Pacific became popular. In England the South Sea Company was formed in 1711, though not to much purpose in the Pacific either then or later. The bitter satirist Jonathan Swift who was pouring out pamphlets and sharp-tongued books throughout the Dampier voyages sent his famous Gulliver on his travels to odd corners of the Pacific. The nasty little Lilliputians lived where the state of Victoria, Australia, now prospers. The friendly, sensible giants of Brobdingnag were the forerunners of the Californians. Other writers, French and English, produced collections of voyages, often lyrical about the alleged value of *Terra Australis Incognita.* One John Campbell 'knew' the missing continent would be found somewhere to the south of the sailing track between Juan Fernandez, off Chile, and Dampier's New Britain. Its discovery, he declared, would open a 'new trade, which must carry off a great quantity of our goods and manufactures. ... It would greatly increase our shipping and seamen, which are the true and natural strength of this country.' Maybe. But the ships and seamen were not sent.

France was also interested, claiming prior right of discovery from some vague land – probably Madagascar – sighted by the Sieur de Gonneville while on an early interloping voyage round Good Hope in the tracks of da Gama. Naming this land Southern India, and carrying off a local prince as proof of his discovery, the French nobleman was intercepted by a privateer on the homeward passage. His journals and papers were seized and apparently lost. Though the prince safely reached France, neither he nor de Gonneville knew where they had been. Equally vague was the 'Cape of the Austral Lands', another French discovery somewhere down south: the astonishing Frenchmen could smell the spices of the East even in sub-Antarctic mists, and they displayed a stubborn courage and forlorn devotion to probing the lonely, useless (except for seals and whales) waters of the far South. Moved by the optimistic raptures of Quiros (whose greatly ambitious hopes his brother-pilot Torres had long disproved) French seamen sought his *Terra Australis* in odd places indeed. If it were so huge as some theorists said it was, its coasts might show up anywhere. One wonders what possible use they would have been in the sub-Antarctic, in the sailing-ship era.

A French scholar and magistrate named Charles de Brosses who never was at sea nor made any sort of voyage at all, was the next major contributor. In 1756 – the year after an obscure James Cook joined the British Navy – he produced his *Histoire des Navigations aux Terres Australes,* which was, one way and another, to have a profound effect on that seaman's career. President (of the parliament of Dijon) de Brosses threw out the sensible suggestion that discovery and development of these southern lands ought not to be left to private companies operating for profit nor sought by conquest, but should be led by States. These *terres Australes*, he reasoned, covered a third of the world. What greater State to bring them all into the modern world than France? The great southern world, he argued, was in three divisions, of which Polynesia and Australasia were in the South Pacific and South Indian oceans, and Magellanica in the South Atlantic. 'Magellanica' was a hangover from Bouvet and the Sieur de Gonneville, and 'Polynesia' was a widespread group of fascinating islands. 'Australasia' was real and contained a continent. It was France's loss that she paid too little attention to the learned de Brosses. His work was stolen by one John Callander in England, who appropriated not just the scholarship but the ideas of empire as well.

In the meantime, along with all this intellectual turmoil, there had been some action. In 1721 the Dutch merchant Jacob Roggeveen, at the age of sixty-two, sailed from the Texel with three small ships with the idea of searching for *Terra Australis*, on his own account, to the west'ard of Cape Horn towards New Zealand – a tough proposition for an elderly gentleman who was no seaman. He stumbled on Easter Island where Davis Land was alleged to be: he lost a ship among the Tuamotuan atolls and nearly lost the other two as well: he saw Samoa, and sailed thence by the beaten track to Java round the north of New Guinea. Arrived at Batavia, his compatriots seized him and his ships as unauthorized interlopers. The old gentleman was hustled back to Holland as a prisoner, which was rather unkind.

About the same time Vitus Bering, a Dane sailing in the service of Tsarist Russia, made a real contribution. Pressing frail ships from Kamchatka into the cold mists of the north, Bering established the fact that Siberia and America were not joined. There was a wide strait between them. Impeded in the south by a long chain of islands strung there like a deliberate obstacle, in the north by drifting Arctic icebergs, and harassed by fogs and mists almost eternally, at least there *was* open water for any greatly fortunate seaman who managed to sail that far round the top of the world. Optimistic atlases had shown a strait there before, of course (just as they showed enormous lands labelled *Terra Australis* and Magellanica). But nobody before Bering really knew.

Bering had found his straits by 1725 and pressed on to see, in time, the noble ice-topped mountain-peaks of Alaska touched to glory by the Arctic sun. His discovery was of worth to Russian fur-traders who moved in among the Eskimo and Indians: no ships came through from Europe either east-about past Siberia or west-about by way of Hudson's Straits. Whether there was a passage for ships by either way Bering did not live to find out,

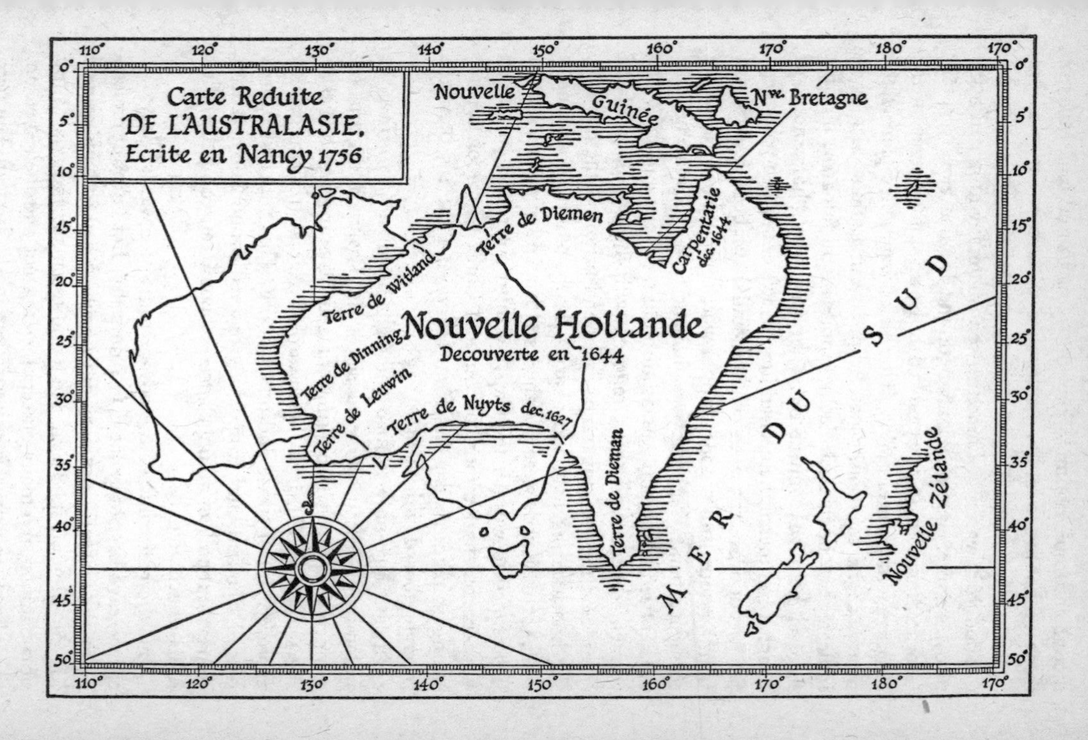
Carte Reduite
DE L'AUSTRALASIE.
Ecrite en Nancy 1756
Nouvelle Guinée
Nw. Bretagne
Terre de Diemen
Carpentarie dec. 1644
Terre de Witland
Nouvelle Hollande
Decouverte en 1644
Terre de Dinning
Terre de Leuwin
Terre de Nuyts dec. 1627
Terre de Dieman
MER DU SUD
Nouvelle Zélande
110°
120°
130°
140°
150°
160°
170°
180°
170°
0°
5°
10°
15°
20°
25°
30°
35°
40°
45°
50°

though incurable optimists kept on trying to find one. Many of them did not live to find out, either.

Though it was not planned to contribute anything to Pacific discovery, the circumnavigation of the English naval captain George Anson in 1740–44 proved to be a stepping-stone in the long story of its exploration. There was war with Spain at the time. Reports to a somnolent Admiralty in London seemed to indicate that the Spanish Pacific empire might fall apart if active English help were given there. In any event, harassment of Spanish commerce there should not be left entirely to privateers. So Anson was sent with what should have been a powerful fighting squadron – four ships-of-war, several store-ships – to torment Spain where she was thought to be most vulnerable. Anson tried to take his squadron round the Horn and the Horn tormented them. His ships, manned with the sweepings of the 'press' and the scourings of barracks and prisons – the maimed, the hopeless, the halt, the weary – badly outfitted, criminally under-provisioned, were thrashed and beaten by violent storms. Two failed to weather the Cape: a third got round only to be wrecked on the lee shore of southern Chile. Crews of the surviving ships died like flies from scurvy and the ills they had brought aboard with them.

At last, able only to man and fight his own *Centurion*, the indomitable Anson staggered across the Pacific, the gaunt 60-gunner weeping her caulking out and dropping her bodies in the sea as she sailed, to the reward of a great treasure galleon captured full of specie and other wealth. This enriched Anson, helped England, and temporarily relieved proud Spain at least in part of the ills of an unbalanced economy caused by the influx of too much gold.

But the real value of the voyage came afterwards. Raised to the peerage later, appointed to Admiralty, Lord Anson remembered the woeful, chaotic conditions which had so handicapped his squadron off the Horn. He found too many navy yards full of corruption, too many officers incompetent and their men undisciplined, and naval administration almost farcical. The need for thorough reform was urgent and obvious: Anson provided capable, thorough administration and restored the Navy's efficiency and morale. From 1745 almost continuously until he died in 1762, he was at Admiralty, much of the time as First Lord. Above all, he chose good, effective men to be his captains and his admirals, and such men served him well.

So the Rodneys, the Howes, Jervis, immortal Nelson began to take over. Under their brilliant kind, the Pallisers, the Saunders, the Colvilles and all the competent rest grew up in the Navy, and served that increasingly efficient service with magnificence rising, at times, to glory.

No Anson, no James Cook – it is as simple as that. For no such lowly warrant officer would have been given the chance to lead great British expeditions and make immortal voyages if discerning, competent officers had not early noted his worth and marked him for advancement when the moment came, in a reformed Navy which valued his services and could use them.

When Lord Anson died James Cook was an unknown surveyor on the coasts of Newfoundland. His chance was to come, when at last England entered the Pacific race, and took up the quest for *Terra Australis* cognita or incognita, to end that shadowy business for all time.

There were others, of course. The first, in 1764 was the Frenchman Louis-Antoine de Bougainville – the same aristocrat who, as a soldier, had broken the British blockade of the St Lawrence and made Wolfe's work so much more difficult there. Turned sailor for the moment – or at any rate, leader of a seagoing expedition under naval discipline, which is perhaps a different thing – Bougainville sailed from a France at last alive to the possibilities of the far Pacific, which included a probable replacement for French Canada. His orders were first to claim and colonize the Falkland Islands as a useful base if any of *Terra Australis* was in the South Atlantic, thence to sail into the Pacific and across it, solving all its mysteries on the way. He came near doing all that, too: at any rate, he touched at Tahiti, re-discovered the Solomons and the New Hebrides, and, for the first time since Torres, refused all nautical advice to turn away in time from the Coral Sea and instead, stood boldly in past the limits of caution. He was not far from the Queensland coast when, alarmed by the labyrinth of reefs and shoals all round him and the increasing difficulty of avoiding them while the south-east trade wind pressed his frigate onward, at last he turned away. One sees his brave *La Boudeuse* foaming along though her sails are shortened, the trade wind balmy and glorious, the deep-blue water frothed to white at the graceful bows. But it is frothy white elsewhere too, much too frequently. For the waters of the Western Pacific here are all too pleasantly attractive to the pernicious and industrious polyp, and coral reefs are strewn with altogether too great a profusion. Many are below the surface and so hard to see. Others shine with the golden sands of minute cays and miniature islands. They are all death to ships.

On the frigate's quarterdeck a group of anxious officers kept careful watch, listening for the noise of seas breaking on coral reefs, watching for the white blink on the horizon ahead which would tell of flung-up breaking water there, listening too for the shouts of the wary lookouts aloft. For this was dangerous sailing, and they knew it.

One wonders, as the frigate twisted and writhed her boldly running way into that most leeward corner of all the reef-strewn Coral Sea, whether the expedition's naturalist, M. de Commerson, stood with them with his good-looking valet in attendance – the same the Tahitians had discerned at once to be a woman, dancing around her with delight shouting *Wahine! Wahine!* for she was the first European woman they had ever seen, and they would have liked to make off with her. Had she removed the coarse canvas bodices and rough sail-cloth stays that had lashed in her femininity? M. de Commerson's messmates had wondered why his valet shared his cabin. She still did.

At last prudence claimed victims. The frigate turned back, to claw her way to wind'ard past the Louisiade Islands and all the eastern end of New Guinea. In due course she made it, and later the long run home again to France, bringing Bougainville back in good time to freshen his military technique to help America in the Wars of the Revolution. The able, pleasant Bougainville left his name on one of the long-lost Isles of Solomon and with a lovely South Seas flower. He possibly had also the distinction of carrying the first adventuress-circumnavigator.

While Bougainville was at the Falkland Islands the English Captain Byron was there too, on the same quest. They met in the Straits of Magellan. On his way through the South Atlantic back to France in 1769, Bougainville caught up with yet another English circumnavigator. This was the much abused and rather badly treated Philip Carteret, lumbering homewards with a lubberly tub misnamed *Swallow* which never should have been sent beyond the English Channel. Like the senior English officer with whom Carteret had originally sailed in 1766, it appears that Bougainville treated Carteret in a somewhat cavalier fashion, too. He had sailed on Carteret's tracks in the Pacific from the Solomons to New Britain and had news of the *Swallow* both at Batavia and the Cape of Good Hope. Yet the French party which pulled across to the stumbling little wallowing waggon passed themselves off as from an East Indiaman, while pumping the English for all they were worth. A French Canadian in the boat told Carteret the truth, and he was annoyed. The French frigate sailed on her way, her wake bubbling, while the deplorable *Swallow* by comparison seemed to stand still in the sea, sulking at her own lugubrious reflection. Poor Carteret! It was then well over two years since Captain Wallis in the fast frigate *Dolphin* had sailed away, leaving his sluggish consort to do the best she could at the Pacific end of the Straits of Magellan – or go back to England.

Why on earth had Admiralty sent her out at all? As some sort of blind? Then why waste the time of a good captain and good officers? Carteret, indeed, though sailing a tub made a far better voyage than either hustling 'Foul-Weather Jack' Byron or the worrying Samuel Wallis. For Commodore the Hon. John Byron was ordered, with astonishing and baseless optimism, to search the South Atalantic for a will-o'-the-wisp known as Pepys Island, and then to sail into the Pacific to find Drake's Bay somewhere vaguely on the coast of California. Using this, when found, as base, he was to find the North-West Passage from the Pacific side and sail back to England through it. If there were no such passage, then he was to continue westwards across the North Pacific to Java or to China as he chose, and home again round the Cape. Admittedly, this was quite a voyage. As an incentive there was an official premium of £20,000 waiting for the first British seaman to sail through Canada by way of any passage he could find between Hudson's Bay and the Pacific Ocean.

Byron never claimed that £20,000. Neither did anybody else, for good reason. But Foul-Weather Jack did not even look. He did go to the Falklands and into the Pacific. He streaked across by the long-established diagonal route as if intent on one thing only, to get back to England as quickly as possible. This he did. In fact, he set up a record for a circumnavigation which stood for many years. Only he was pleased.

There was an excuse for Byron, who was perhaps an unfortunate appointment. He had been wrecked down near the Horn out of the *Wager* – one of the Anson ships – as a young midshipman and was fortunate to survive. It took him five years then to get back to England, after dreadful hardships on the coast of south Chile, a searing experience he never forgot. The fine coppered frigate *Dolphin*, with her 190 men and good equipment including a 'machine' for purifying water, was wasted on him.

His next appointment was as governor of Newfoundland: after that, for a while, he had an indifferent squadron based on Sandy Hook.

The poet Lord Byron was his grandson.

Byron's poor results could not be accepted, especially when it was known that Bougainville had done much better. The *Dolphin* was given a quick refit and sent off again with a new captain named Wallis with orders this time to find *Terra Australis*. Wallis did no such thing. He didn't look, either. He did stumble upon the lovely island of Tahiti, largely because he was sailing westwards in more or less the same latitude and Tahiti is a high, volcanic island which on a clear day can be seen from the masts of a sailing-ship seventy or eighty miles away. *One* of these hustling circumnavigators had to find something. It was odd, indeed, how so many had kept on missing so much.

It is wrong, of course, for later generations to blame seamen sent to the eighteenth-century Pacific for what they failed to do. After all, they were on their own in frail ships trying to cross an ocean of appalling magnitude. First they had to weather the worst cape of the whole sailing-ship era, at the end of a long passage out from Europe. It was either fight past the Horn or take on the miserable Straits of Magellan, which was worse. To fight a square-rigged ship *into* strong winds and gales is a vastly different thing from running with them, or before them. The ship bucks, ships water, strains her hull, her masts and yards, her cordage rigging. Even in the mid eighteenth century, frigates like the *Dolphin* and the *Boudeuse* were strung together aloft with what amounted to little better than a complicated and vulnerable structure of sticks and string – thin sticks and thick string. The chafe was appalling. Shelter and heat they had none, for the gunported 'tween decks and the stern-windowed great cabins were draughty, wet, and cheerless.

These ships-of-war were hangovers from the days when such vessels normally did most of their sailing in the good-weather months (which in northern Europe could be bad enough) and were never expected to go on slogging it out at sea month after month for years. Their cordage rigging, wooden yards, rope-rigged steering and so on were not designed to go on and on taking the sea's and the sea wind's punishment month after month, year after year. They needed repair facilities, dockyards, rope-walks. They needed fresh biscuit and fresh water, too: only big, special bakeries could turn out biscuit enough (they called it bread) for the scores and scores of men they had to carry, to be sure of humans enough to fight the ship and survive to sail her when the scurvy struck.

There was no dockyard in the whole of the Pacific and Indian Ocean areas. After the English Channel, for them Batavia (Djakarta) in Java was the next real base, and that was Dutch. There was no spar-wood or good mast timber growing in the South Seas islands anywhere known before the Cook voyages; the islanders, like the Asian seamen, knew how to twist cordage from coconut fibre, but in short lengths only, fit for their canoes. It was some time before Europeans thought of improving on

Polynesian skills. As for biscuit, their English stores would crumble to a befouled dust before there was hope of replenishment, and their indifferent salt meat harden into a putrid, useless mess, if it hadn't come aboard that way in the first place.

No, European ships with European stores, crews, and maintenance problems turned their shapely sterns on all hope of succour when they beat past Cape Horn.

These English naval frigates had another problem, even worse. The career officers commanding them, brought up in the Navy from boyhood, had nothing to expect from the arduous duty of a Pacific voyage – no prize-money, no glory. For this was a quiet warfare of little appeal to the service as a whole, with an excellent chance of making some fatal discovery with the ship (or ships) themselves, of being overwhelmed off the Horn or bashed to pieces on some horrible lee shore in the Straits of Magellan if they went that way, or ripped apart on one of the far too numerous mid-ocean coral reefs. There were plenty of them. No wonder the commanding officers all made for that 'comfortable' old diagonal route through the trade winds and round the north of New Guinea. A circumnavigating voyage of discovery was a trial which it was sufficient to survive.

The remarkable thing is that any of these ships did survive their voyages at all.

Byron and Wallis had the coppered *Dolphin*, which could be expected to keep out the boring teredo worm and hold back the speed-killing barnacle. But copper sheathing was then experimental and worrying. The inter-action of iron fastenings and copper plates was imperfectly understood. Byron feared that the *Dolphin's* rudder might drop off almost at any time, for her iron pintles wore down to needle thinness. A rudderless ship can perhaps be handled, but hardly as a vehicle for discovery. Like the rest, Byron had scurvy to contend with, too. And what nightmares must have disturbed such sleep as came to him, from that ghastly wreck of the *Wager*! Day and night the floundering rudder, below the stern windows, went thump-thump-thump on its loosened pintles in the sea, shaking the whole stern structure; and any thump, he knew, could be its last. There was no dock to fix it. The *Dolphin* was not a ship to be hove down on a coral island.

Magellan, Quiros, Torres, Tasman, Schouten and le Maire, Roggeveen, Drake, Dampier, Bougainville, Byron, Wallis, Carteret had come and gone. After the lot of them half the Pacific remained quite unseen and wholly unexplored. Tasman's New Zealand could extend southwards to the Pole and eastwards to Easter Island or 'Davis Land', and the east coast of New Holland reach Tasmania and almost to the Fiji Islands, for all anyone knew. As for *Terra Australis Incognita*, the sum of all the voyages had not shown if the place existed or not. All they had done was fail to find it. There was plenty of area in the Pacific still unseen where a great continent could be found.

By early 1768, there was increasingly urgent reason – they both thought – for France and England to get on with finding it. The French, as a Power, had been thrown out of North America, cut back to size in the West Indies, and more than threatened in India. The English were soon to be thrown out of America, too, by their own rebellious colonists. Already the rumblings of this could be heard.

The nation which first found the allegedly great new continent and its imaginary riches could, in time, use it for the domination not just of the Pacific, but the Indian Ocean and Asian waters as well. It was a political prize of the first importance and – it was hoped – it should prove a commercial bonanza too.

But who was going to find it? Naval expeditions cost large sums: naval captains became dictatorial lawmakers unto themselves once out of sight. It seemed useless to fit out more frigates (or the same old *Dolphin*) for more career captains to sail round the world, carefully avoiding the area where the missing continent was most likely to be found.

At this stage, there appeared publicly on the scene the astonishing figure of the Scot Alexander Dalrymple, hydrographer, seafarer (of brief experience), Fellow of the Royal Society, indefatigable researcher into the Pacific story and collector of Pacific voyages, and doctrinaire extraordinary who could easily have wrecked all Britain's hopes in the great South Sea, and nearly did. For Alexander Dalrymple took unto himself vested rights not only in all matters pertaining to *Terra Australis* incognita or cognita from the academic point of view, but was determined to be its God-given discoverer – the Columbus of the Pacific Sea! Mr Dalrymple was an outstanding example of that unfortunate but still prevalent type of scholar who, having once construed the available evidence on anything into a theory to his own satisfaction, thenceforth and for ever afterwards can see only evidence supporting that theory and confounding all others. For him, *Terra Australis* was by no means a vague theory. Its coasts and mountains had been glimpsed a dozen times by discoverers unable, for one reason or another, to follow up. From the coasts of New Zealand in the west to 'Davis Land' allegedly not far from Chile in the east over 4,000 miles away, from Quiros's Espiritu Santo and Mendana's Solomons in the north to the Antarctic in the south – with tropics, temperate zone, frigid zone, untold thousands of square miles of each – his *Terra Australis* was quite a country. It had been glimpsed fleetingly but never investigated because all the seamen, for their own foolish reasons (as Dalrymple saw them), had stuck to the trans-Pacific diagonal route. Nobody had sailed where it was.

Oddly enough, New Holland, the real *Terra Australis*, did not enter into Dalrymple's calculations at all, though he was one of the few men who was aware that the Portuguese Torres *had* seen its north-eastern coast and *had* sailed through between its northern tip and New Guinea. For Dalrymple had come upon documentary evidence of the Torres voyage, found in Manila. He kept this to himself, though he mentioned it in a slim book with the long title *An Account of the Discoveries Made in the South Pacific Ocean Previous to 1764*. This book he had printed in 1767 but not published then.

He knew, early in 1768, that the Lords of Admiralty were planning yet another Pacific voyage, to find and take possession of *Terra Australis* come what may. The Royal Society had requested a ship to go to the South Seas as a convenient spot to observe a transit of Venus for astronomical and navigational purposes, and this was good 'cover' for Admiralty's real intention. This was to sail westwards and south-west, and to lay bare the secrets of the South Pacific where they were to be found.

Who should command this ship and expedition but he, Alexander Dalrymple, F.R.S., God's chosen for that very purpose? Naval captains, buccaneers, privateers, seekers after business gain and visionaries all had had their chance, and failed. What naval captain was available to do better now?

Dalrymple, whose family was ancient and honoured in Scotland where they had long held the earldom of Stair – an older brother was Lord Hailes, a noted judge and humanitarian, friend of Dr Johnson – had on the face of it, every prospect of receiving the appointment he coveted. After all, no one else wanted it. Few knew of it. In the Navy, when it was whispered about, no one volunteered. There was too good a prospect of profitable war and career captains wanted to be in on that.

Earlier in the eighteenth century, the appointment of the dogmatic, overbearing Dalrymple would have been assured. It seemed reasonably assured even then, not only to him. But there were Lord Anson's well chosen men in key posts at the Admiralty. The fruits of Lord Anson's policies were still obvious there. Some of the bright young men he had marked for promotion were in key posts, and they were unlikely, when it came to appointments in command of difficult and important seafaring enterprises, to be bothered much by family influence if other matters weighed more. There had been mistakes enough. This time England had need of a thoroughly capable officer who would carry out his orders, and to do what he was sent to do.

Well, maybe there *was* a man in the Navy of that stamp – an obscure warrant officer who had shown himself to be a skilful seaman and surveyor, mathematician, astronomer, and excellent cartographer as well – a quiet, tall man with a marked Yorkshire accent, striking in appearance, noted for diligence, adaptability, and complete trustworthiness. These qualities had already brought him to the notice of senior officers though after over twelve years in the Navy he was not even a commissioned officer, and he had no family or other influence at all. He was forty years old.

The name was Cook – James Cook, a former Whitby seaman.

Alexander Dalrymple was far from unique in never having heard of him.

CHAPTER FIVE

A Coal 'Cat' Sails for Cape Horn

THERE are few academically provable facts about the selection of James Cook to command of H.M.S. *Endeavour*. The records are so incomplete and vague – or non-existent – that it is reasonable to assume that, at the time, the matter was at first hushed up and the negotiations carried on verbally and very quietly until all was settled. It is too much to accept that so unorthodox an eighteenth-century English naval officer and so thoroughly unorthodox an H.M. ship as the 'bark' *Endeavour* just chanced to step out of the blue together, from the same home port. The Navy was not in the habit of acquiring North Sea collier barks or barques as King's ships. The highest a coal cat might aspire to, normally, was temporary charter as a transport for use with her merchant service crew in home waters, in time of war.

As for Cook, it was over ten years since Palliser had more or less promised him lieutenant's rank – the lowest commissioned status then – after six years' service, but he had not been so promoted. This probably suited Cook quite well. He was better off financially as Master. He had an independent command and ten shillings a day, which was more than lieutenants were paid. It was regular, too: there was no 'half-pay' about it. He was employed continuously at varied and useful work he obviously enjoyed, which gave him considerable satisfactions. His pay kept the modest home he had set up in the Mile End Road in London going very nicely. Indeed, Mr James Cook, Master and Engineer R.N., some time in command of the schooner *Grenville*, full-time surveyor and cartographer (his charts were already becoming available at the chart publishers in London, for the Navy had not then a department to produce them) had already come a long way, as a professional ex-merchant seaman.

There were many lieutenants: a Master of his experience and abilities was rare, and a Master known favourably in high quarters even rarer.

His abilities and reputation were known only to the knowledgeable few, who were not loquacious. He had come to notice above all as a quietly competent seaman, able to rise to problems and somehow produce or achieve something more than the expected. But for the quietly competent no flags fly and preferment comes slowly. When the savants of the Royal Society petitioned Admiralty for yet another ship to be sent to the Pacific, they assumed that their member Mr Alexander Dalrymple would be the principal observer, which was reasonable. They also expected that he would be appointed to the command, which was not. He was not a professional seaman. There was precedent for the selection of a socially acceptable leader, preferably aristocratic, who was not a seaman. Few of the great pioneering discoverers were just seamen, or career seamen in any real sense at all. Leadership of pioneering voyages of discovery usually had meant, first of all, the ability to organize them. To do this required influence, the right contacts, personal presence at court and in the chambers of the mighty. Da Gama, Magellan, Cabral, the Cortereals – all the great Latin pioneers – were *hidalgos*. Whatever Columbus was, he was not a seaman. Expeditions by sea necessarily carried sailing Masters, pilots, boatswains and the like who knew their business and practised it for the lack of competence at any other: but overall command was not for them. Their professional voices were heard in the ships' councils.

In this case, the original driving force behind the expedition was the Royal Society's, and there was precedent for one of its Fellows to be given temporary command of a King's ship. In 1698, Edmund Halley, sometime Astronomer Royal, Oxford don, secretary of the Royal Society and editor of its *Transactions*, mathematician, and astronomer extraordinary, was commissioned as a temporary Captain, Royal Navy, to command a King's ship on a two-year expedition. This was to carry out astronomical and magnetic research at sea, mainly in the southern hemisphere, which Halley was particularly fitted to do.

It was also true that the ship he commanded, the *Paramour Pink* ('pink' was a term for a ship of peculiar hull-type, just as 'cat' was) had been financed and built by a patron of Halley's and presented to the Navy for this use. Apart from Halley, she was navy-manned. She had not sailed far before her first lieutenant, a career officer named Edward Harrison, was causing trouble. Among other valuable activities, Halley was working at the problem of finding longitude at sea. His solution was to perfect the working of 'lunars', a series of complicated problems based in part on the occultations of fixed stars behind the moon. Using Mars and the moon, he had worked out the longitude of Oxford. One trouble in this method was that the occultations did not occur to seamen's need but by their own laws. An observer ashore could abide by these. Another lunar method was to observe the distance between the moon and a star, on the theory that the motion of the moon past fixed stars is considerable and can be tabulated to provide a skilled observer with data to assist in the solution of astronomical problems. In the absence of a better method, lunars seemed to offer a solution of the longitude problem. If anyone could make them work, then it was Halley.

The first lieutenant of the *Paramour Pink*, it seems, had a complicated theory of his own on such matters with which he was well satisfied. He explained it to Halley. It was nonsense. He explained it again. This became boring, and Halley did not suffer fools gladly. He probably said so. Harrison became insubordinate to the point of mutiny. Captain Halley put him under arrest, sailed the *Paramour Pink* back to England, court-martialled his first lieutenant, won the case, and sailed again to prosecute the voyage successfully without a naval lieutenant at all. The Navy was rather oddly run at the time, perhaps, but if this was mutiny, it was scarcely Halley's.

Alexander Dalrymple was a clever man and a scholar, but he was no Edmund Halley. Nobody gave him a pink: nobody meant to give him a cat either, despite his influence. Dalrymple was also known through his service with the East India Company, in management, at sea (locally in eastern waters, and very little) and as hydrographer and acting diplomat.* As this last, he had negotiated a successful treaty with the Sultan of Sulu. If *Terra Australis* existed, it was reasonable to expect him to be able to negotiate with its potentates, if any.

Unfortunately, he was also an irascible, dogmatic and utterly uncompromising eccentric, who suffered increasingly from gout. These qualities could make for a certain lack of harmony in the close quarters of a small ship. It might have occurred to one of those eighteenth-century minds that though the East India Company had dismissed him, this *Terra Australis* could be an immense prize and its discovery a temptation. What was to stop a non-Service commander, if he found it, dodging off to India first instead of sailing home to give the news to George III? Dalyrymple's behaviour gave inadequate evidence of his general stability.

Dalrymple had made himself the outstanding scholar in Britain if not in Europe on the general matter of the theoretical southern continent. He had committed himself to the idea of is existence, for which he had argued strongly in a privately published work entitled *An Historical Collection of the Several Voyages and Discoveries in the South Pacifick Ocean*.† But academic insistence on the shape of the world was no qualification for leadership of an expedition to establish just that. A discoverer's first business was to find what was, not search for what he imagined should be. As for that, so was a scholar's. Negotiations between the Royal Society and Admiralty went on

*As for his connexion with the Honourable East India Company, it is significant that – in his own words – he had been 'capriciously dismissed from the Company's Service, without any reason assigned'. Apparently he was being pig-headed at the time about an idea of his to exploit a grant of the island at Balambangan or Balembangan, in the Balabac Straits between North Borneo and Palawan in the Philippines. He produced a dogmatic work about this, too – *An Account of What Has Passed Between the Indian Directors and Alexander Dalrymple*, etc. (London, privately printed 1769: sold by J. Nourse.) He wanted the Company to change its whole plan of business and make Balambangan the centre of British trade in the Far East, with himself in charge. The Company refused. He argued. The Company fired him.

†My two-volume edition was printed for Dalrymple and on sale at J. Nourse, bookseller to the King, and two other London booksellers.

for some time, at first favourable to the Scotsman. But in the end, he spoiled his own chances.

It was at last made clear to him by Lord Hawke that though he would be welcome in the ship sent, it must be as a civilian observer serving under the command of a naval officer.

Command or nothing! declared the scholar, very sure of himself.

No one but a serving naval officer bred in the Navy shall have command, said Hawke, very quietly and firmly, sure of himself too.

Dalrymple could pull strings. Indeed he could haul on strong ropes.

He did.

Nothing happened, to his considerable astonishment. The eighteenth century was a string-pulling one in England, in which it was usually almost impossible for influence properly applied not to achieve its object. If men might for the moment fail, there was no shortage of strong-minded, unscrupulous and well-placed women, young and old, to take over. Doubtless they did in this case. For once, they all failed. So did the Royal Society which perhaps, knowing Dalrymple and Hawke, was not greatly surprised.

Dalrymple at last withdrew in a dudgeon. If the Secretary of Admiralty meant what he said, declaring that he would sooner lose his right hand than see a non-naval officer again command a naval ship, the Scots theorist meant what he said, too. Perhaps what really upset him was the thought that someone else would have the honour and glory of going down to posterity as the Columbus of the South Seas.

In a way, it is perhaps a pity that the gouty, doughty Dalrymple was not sent off in some ship – preferably Carteret's tub *Swallow* – to look for his beloved continent. Standing for ever westwards in the far south where the west winds blow until he was the last man left alive, while the ship disintegrated into that landless abode of screaming storms and the cynical, cold-eyed albatross – here was the stuff of legend! The Flying Dutchman would be a fairy king beside the dour non-flying Scot – callous, confident and implacable, driving his ship on endlessly towards every cloud-wraith or mist-line that any optimist or bleary-eyed pessimist, drunk or sober, had ever 'sighted' and listed as land in the great South Sea.

James Cook knew nothing of the machinations going on round command of an expedition he had probably not then heard of. He led a full life, happy in his appointment and conditions of service. Plans for the real purposes of the coming voyage were secret, though the Royal Society's wish for a King's ship to make a South Seas observation of the transit of the planet Venus was not. The Society had been considering this for some time. Many years earlier, Halley had forecast the transit in 1769, with no other afterwards for a hundred years. Data on the transit was essential for the advancement of nautical science and astronomy, in order, among other things, to calculate the earth's distance from the sun, then unknown. One of the places most suitable for a series of observations was the Marquesas Islands in the South Pacific.

All this was very interesting, and Cook had shown himself an interested student of nautical astronomy. But as far as the records show, there was no reason for him to dream of association with the enterprise. Apart from a few Atlantic crossings, then commonplace, he had never made a long voyage. He had commanded nothing but a small and undistinguished schooner on isolated (though demanding) service. At the best, he was a competent nonentity. The theory has been put forward that Admiralty's intention originally may have been to have Cook go as Master of the *Endeavour* under a gentleman from the Royal Society, in overall charge (like Bougainville earlier for France). But Dalrymple was so impossible that he forced Lord Hawke – a strong-minded and very able man no more content to suffer fools gladly than Halley – to think again. Once the idea of Cook in command occurred to Admiralty, the problem was solved. The Whitby seaman had shown a remarkable capacity for carrying out orders thoroughly and *growing into* his appointments. All his commanding officers and all the senior officers who knew him spoke well of him.

As early as November of 1767, the Royal Society had set up a committee to organize the expedition. A few weeks later, their recommended candidate Dalrymple was observing that he could 'have no thought of undertaking the voyage as a passenger going out to make the observations or on any other footing than that of having the management of the ship intended for the service'. At the end of February 1768, the King requested a ship from his Navy. A week later, Cook is reported as calling at Admiralty, to hand in his journals from the previous year's work aboard the *Grenville* – a normal procedure. As far as official records go, he went back to prepare the schooner for her next voyage to Newfoundland: but a successor was already in mind then if not appointed.

On 1 April of that year, Admiralty informed the Royal Society that the Whitby cat *Earl of Pembroke* had been acquired for the coming voyage, and renamed *Endeavour*. On May 25, over seven weeks later, having passed the examination, Cook was promoted lieutenant, and given command. By then, still a Master, he had been introduced to the Council of the Royal Society and accepted as one of its two official observers, the other being an astronomer named Green from the Observatory at Greenwich. The name of Dalrymple was not mentioned.

As for the *Endeavour*, a seaman is forced to conclude that, at the time, only Cook would have selected such a ship. The pugnacious Dalrymple claimed later that he chose her, of course – 'the ship I had chosen for the voyage', he wrote in a diatribe published in 1773 as *A Letter From Mr Dalrymple to Dr Hawkesworth*, etc., still lamenting his loss of command. The outstanding quality in a cat was her ugliness. If beauty is in the eye of the beholder, it had to be a fond, trained, and experienced eye indeed which perceived the sea-grace and excellence of so homely a vessel. The fat, uncompromising bow, the graceless thrust of the straight cutwater, the austerity of masts and rigging, the stubby work-horse of a hull combined (in a harbour containing Indiamen, frigates, swift sloops and rakish brigs) to indicate, at first sight, a lowly vessel, a plough-hauler among the race-horses.

We find the Navy Board, after its ship-surveyors had looked at a sloop named the *Tryol* at Deptford and another named *Rose* at Sheerness, suggesting on 21 March the 'choice of a cat-built vessel for the said service', pointing out that 'their kind are roomly and will afford the advantage of stowing and carrying a large quantity of provisions, so necessary on such voyages'. If such a vessel were decided on, they added that there were several of around 350 tons lying in the Thames. Admiralty came right back the same day with the formal go-ahead, obviously previously arranged. Within days, the best of the newer Whitby cats in the river had been surveyed and the *Earl of Pembroke* chosen and bought, the price being £2,840 10s. 11d. She was of 368 tons, some 98 feet long on the lower deck with a keel of 81 feet, a maximum beam of 29 feet 3 inches and depth in hold of 11 feet 4 inches. The surveyors described her as a 'square stern bark, single bottom, full built' three years and nine months old, 'a promising ship for sailing of this kind ...'

Who detected her promise and aided those skilled old master-shipwright and master-rigger surveyors in their choice? The answer seems obvious.

Writing later, Cook himself explained the merits of Whitby ships like the *Endeavour* (and later the *Resolution, Discovery,* and *Adventure*), for voyages of discovery. They were strong with comparatively shallow draught, he said, yet good carriers with plenty of room for men and stores. Such ships, in construction, 'were of the safest kind, in which officers may, with the least hazard, venture upon a strange coast' or wander by day and night across the sea looking for such coasts. With their shallow draught they could safely sail near enough to the land with time to turn away from warning sights, smells, or sounds: if at the worst they took the ground, they could sit on it a while without much fear of fatal capsize. Nor were they too large to be beached and repaired, if such a necessity arose.

'These properties are not to be found in ships of war of forty guns, nor in frigates, nor in East India Company's ships, nor in the large three-decked West India ships nor indeed in any other but North-Country-built ships such as are built for the coal trade ...'

Two further advantages Cook does not bother to mention as obvious to his fellow seamen. These colliers were comparatively lightly rigged and simple to handle. There was no very heavy gear in them, as in larger ships. Further, he was thoroughly used to them.

The *Earl of Pembroke* was taken to H.M. Dockyard at Deptford, on the Thames not far from Greenwich, where the king's ships had been built and refitted since Henry VIII built his *Great Harry* there in 1514. She was nothing but a strongly built cargo-carrier with rough accommodation for minimum crew. Now she had to carry a crew of at least seventy, as well as observers, scientists, artists and servants and so forth, and provisions, stores and liquor for the lot to last two years. She had to have enough guns to take care of herself, five new anchors, and three large boats – longboat, pinnace and yawl. She had to be protected against the borer-worms of warm seas. These borers do not just cut into a plank and eat through, which is bad enough. They insert themselves into all the underwater body of a ship, plank by plank, and then devote their lives to eating happily *along*, inside the plank, breeding more of their kind to do likewise until in the end, though still retaining the outward appearance of a ship, the hull is fit to fall to pieces.

Roman ships were sometimes sheathed with lead, Arabs with coatings of lime mixed with camel-fat applied once a quarter, European ships with a skin of extra planking. Between the outside skin and the hull proper, a coating of teredo-deterrents was packed – oakum and tar and so forth. In 1768, copper sheathing was being tried experimentally. Cook took the view that copper plates were too easily damaged by groundings, chafe from boats working alongside, anchors, and so forth, and the risk of infection from the borers was too great on long voyages. Therefore he had the *Endeavour* given an additional skin of light planking over the usual deterrents, nailed on with thousands of flat-headed nails which were driven home as closely as possible. Whatever else the little cat was to be, it was not a portable meal for the South Sea borers.

There is documentation for all this. With their customary

thoroughness, the shipwrights at the royal yard took off the ship's lines and measurements, and prepared drawings and plans, with their own alterations shown additionally. Unfortunately, this did not include the masts, yards, and rigging. These had mostly to be renewed, and the dimensions of the masts and spars are preserved. The replacement of rigging was done from coils of the right sizes of rope, obtained from naval ropewalks as required by riggers who knew from tradition and handed-down skills exactly what they were doing. Plans were of no aid and therefore not necessary.

A bark's rigging was simple. Drawings of cats show them rigged as small ships, though incomplete on the mizzen. A 'ship' was always – and still is – a sailing-ship square-rigged on three masts (or four or five, if she had that many). In later days, to rate as square-rigged, the wooden masts had to be in three sections, lower-mast, topmast, and topgallant, with at least one sail from each. Steel lower and topmasts could be combined but were rigged the same. The *Endeavour* had a short mizzen consisting of lower-mast and topmast only, setting a small square tops'l, sheeted to the crojack's yardarms and hoisted on its own yard, above a loose-footed spanker. She was referred to as the bark *Endeavour*, or the *Endeavour* bark or Bark: she was not a barque (which is fore-and-aft rigged only on the after mast) and she was not strictly a ship either.

'Bark (*barca*, low Lat.), a general name given to small ships: it is however peculiarly appropriated by seamen to those which carry three masts without a mizzen topsail,' Falconer lays down in his *Universal Dictionary of the Marine*,* adding that coal-trade seamen apply the term generally to 'broad-sterned ships which carry no ornamental figure on the stem or prow'.

So the *Endeavour* Bark, or the bark *Endeavour* was a sort of ship-rigged cat. As for eighteenth-century ships in general, Falconer simply describes them as the 'first rank of vessels which are navigated on the ocean', and admits the term ship to be of 'very vague and indiscriminate acceptation'.

It still is.

* My copy is dated 1780.

Sail plan of the Endeavour

All sorts of things had to be done to the *Endeavour* – gun-ports cut into the hull for the few larger guns; a new deck built in the hold (previously all cargo space) to provide slinging room for the seamen's hammocks; the quarter-deck moved up and carried further forward to provide head-room; cabins, store-rooms and magazines added; the galley (a sort of movable box on deck with a casual fire in it served for a Whitby crew) moved below and extended; more and stronger hand-pumps fitted, stowages on deck and tackles from the spars and rigging provided for three large boats (but no davits, of course); the cargo hatches converted into companionways and fitted with gratings for better ventilation; the capstan, windlass and ground tackle overhauled and stowages provided for additional anchors, cables, and hemp to make more cables; large sail-lockers and a carpenters' workshop, a better steering compass in a better binnacle, eight tons of iron for ballast stowed below, and a heavy ringbolt driven into the stem-head to provide a better gammoning (lashing) for the bowsprit. While all this went on, decks and upper works were caulked, the whole of the masts, spars, standing and running rigging sent down or unshipped, examined, made good or replaced, and installed again. New sails were cut and sewn.

This was a thorough refit, as it had to be when the ship was to be away from base for so long. As a Whitby collier, the smallest of great cabins with a sleeping cabin or two opening off sufficed aft, and a hutch below decks in the eyes for'ard for a fo'c's'l for the men. Her sail locker then was at Whitby, and she was stored for each voyage. Now ten new cabins were hammered into her on two levels aft, as well as pantries and a large lobby off which the upper deck cabins opened, and a small one for the great cabin. No wash-places or lavatories were provided apart from a minute 'heads' in the for'ard corners of the great cabin, accessible from the captain's and the first lieutenant's sleeping quarters – captain to port, first lieutenant to starboard, which was the reverse of the traditional arrangement. It was slightly easier to reach the quarter-deck from the port cabin, and that may be why Cook chose it. With a crew of seventy there was obviously not going to be any spare room aboard, not for anybody.

While all these preparations were coming to fruition, Cook (who had hoisted his pendant and taken command in the basin at Deptford Yard on 27 May) was informed by the secretary to the Admiralty that he was to receive on board 'Joseph Banks Esq. and his Suite comprising eight persons with their Baggage, bearing them as Supernumeraries for Victuals only, and Victualling them as the Bark's Company during their Continuance on board'. Another eight to crowd aboard, several of them demanding single-berth cabins and all demanding something. Where could they all be stowed, to say nothing of their mounds of baggage? This included botanists', artists', and naturalists' equipment to last two years – cases of jars for specimens, pots and presses for plants; canvases, easels, brushes, paints and so on on the same scale; writing materials; books, trinkets for 'savages', nets, trawls, drags and hooks for fishing;

'machines' for catching and preserving insects; spirits to preserve small animals, wax and salts to do the same for seeds: and two large greyhounds.

The party included a Swedish botanist named Daniel Karl Solander, 'well-loved' pupil of Linnaeus, an outstanding scientist with some medical training; another Swede or Swedish Finn, also a naturalist, named Hermann Diedrich Spöring, the son of the professor of medicine at the Finnish University of Åbo (or Turku, as it is now known) who had trained in surgery at Stockholm himself where he had acquired a certain manual dexterity useful in watch-repairing (he brought a set of watch-making tools aboard the *Endeavour* where he came as a sort of writer-clerk-second draughtsman-watchmaker etc. to Mr Banks); a botanist and draughtsman named Sydney Parkinson, a Scot from Edinburgh and an excellent man; another Scots artist-draughtsman named Alexander Buchan who, unfortunately, was an epileptic; two young fellows named Briscoe and Roberts from the Banks estates in Lincolnshire, engaged as 'footmen' – the first (and possibly only) footmen to beat to windward past Cape Horn – and two coloured servants named Richmond and Dalton or Dorlton. This pair, unfortunately, did not weather the Horn, for they died of exposure on a Banks shore excursion on Tierra del Fuego.

Well, there they were, in due course, their bags and baggage dumped on deck, filling the boats, covering the hatches, in the way everywhere, and the dogs tethered to the windlass, howling. The footmen and the coloured lads could sling with the mariners somewhere and take pot luck; but the others were a problem. Where could they find adequate cabins in so small a ship? Mr Green the astronomer had to be accommodated, too. The problem was solved by shifting the ship's own officers – First Lieutenant Zachariah Hicks, Second Lieutenant John Gore (who had been with both Byron and Wallis), Master Robert Molyneux, Surgeon Wm Brougham Monkhouse, Gunner Forwood, with Orton the captain's clerk – down to stuffy, poorly ventilated cabins aft on the lower deck, where it would be hellish in the tropics and grim off the Horn. Mr Banks was allotted a good cabin on the starboard side off the great cabin –

best cabin aboard after Cook's own, which was slightly larger – and Mr Green, Dr Solander, Spöring, Parkinson and Buchan spread round the ship's officers' cabins, which were reasonably lit and could be ventilated, at least a little. The ship's petty officers – the carpenter and boatswain – lived in minute cabins on the lower deck for'ard, and all the rest slung their hammocks or stretched out on the lower deck the best way they could. Once the ship was at sea, they would never all be trying to sleep down there together because of the watch system, and they had precious little personal gear to stow. The lower hold was full of barrels, casks, provisions of all sorts, and stores, all carefully stowed, jammed and chocked off against violent motion. Where the ship's three midshipmen managed to set up a mess one does not know, but it is certain that they managed. Naval ingenuity is great, and midshipmen have always been outstanding in this matter. Perhaps they contrived to set up some hell-hole on the lower deck by the gunner's or clerk Orton's cabins.

The descent of Mr Banks and his party aboard that already crowded cat, though giving the expedition, previously concerned with astronomy and geography, a fantastic group of learned and cheerful young fellows off on Banks' idea for the first 'grand tour' of the world and aiming to contribute greatly to the voyage, added to Cook's difficulties in more ways than one. On the face of it, they could not help but be a considerable imposition on him and his officers. The great cabin – that windowed and graceful spaciousness in the stern of the ship where alone she could be both properly lit (by daylight, and at night by candles swinging in their weighted sockets of polished brass) and ventilated (by the sea air coming in the stern windows and perhaps also a skylight) – was the only retreat a captain had, the only place where he could do anything. It was chartroom, messroom, recreation room, library, writing-room, everything. Its large fixed table and comfortable settees offered all the comfort to be found, its leaded lockers and strong cupboards all the spare stowage. In a 100-foot cat it was small enough, for cats were slim-sterned.

Now Cook, carrying the burden of command on a vital

voyage, was to be pretty well crowded out of it: Mr Banks set his horde to work there (to some purpose), for it had to be his workshop and writing-room (the industrious Mr Spöring doing the writing) and home, too. Banks was a cheerful, exuberant – and arrogant, when he chose – twenty-four; Cook a taciturn, sometimes dour Scots-Yorkshire seaman aged forty, which was old then.

The great cabin of the little *Endeavour* made room for the most fantastic band of circumnavigating young university men – each brilliant in his own way – assembled up to that time for such a purpose, or anything like it. It was unique and wonderful. Indeed, it was incredible. It was also quite an imposition on the generosity of Captain Cook. One begins to understand, perhaps, at least one reason why he was appointed. It is improbable that any other naval captain would have had his traditional quarters so crowded at the outset of so long a voyage. A good sailing-ship commander carries *all* the strain of voyage-making, which is often considerable and always trying. (One speaks with a little experience.) If he can have no privacy ever, save when briefly asleep (with one eye and both ears open) then it might become impossible.

Joseph Banks, F.R.S., was an extraordinary young man, even in the eighteenth century. Son of a wealthy Lincolnshire medical man who became sheriff of his county and an M.P., he was a Lincolnshire landowner of considerable substance, with an inherited income of £6,000 a year, tax free. (A gross income of £100,000 a year would not give him anything like the same substance in the 1960s.) A skilled steward looked after the land: Mr Banks had a house in London as well as the family seat of Revesby Abbey near Boston in Lincolnshire. His education was probably unique. He was an 'old boy' of both Eton and Harrow and a graduate of Oxford, where he was admitted at a 'gentleman commoner' at Christ Church in 1760. His scholarly diligence was not great, but his peregrinations between famous public schools followed his own inclinations and not masters' directions. He might well have been at Cambridge, rather than Oxford, for going up to read botany as a subject allegedly taught at Oxford, he found a professor but no instruction – no lectures, no tutorials, no guidance whatever. In the eighteenth century (and later) this might have been acceptable, for a gentleman commoner was not expected to be a scholar and could please himself more or less what he did or neglected to do. Banks was different. He *meant* to read botany. Discovering a young man who was described as a lecturer in Botany at the other place, he obtained permission from his somnolent professor to import him from Cambridge forthwith, at his own expense of course (including salary, doubtless not large) and so, with the arrival by coach of Mr Israel Lyons, the subject was actually offered at Oxford and not just listed. Mr Lyons had really been at Cambridge to teach Hebrew, but there were others to do this; and he was also a botanist and astronomer. The astronomy he could keep for his personal enrichment, for the young Banks (and later the old) was no mathematician of any sort, but he had been a sincere and ardent botanist since early youth. Mr Lyons was an excellent tutor: the pair prospered in their chosen subject.

Mr Banks did not forget his tutor. When in 1773 his Oxford friend Phipps, by that time a captain in the Royal Navy, wanted an astronomer to sail with him on yet another difficult Arctic voyage in hopeful quest of the North-West Passage and the North Pole as well, Israel Lyons went by Banks's recommendation. By this arrangement Captain Phipps was well served: whether Mr Lyons was one cannot say.

A Christ Church man had influential friends everywhere, but it is odd that there is little tradition of Banks at Oxford today. The former Deputy Librarian of Christ Church, the late Mr W. G. Hiscock,* showed me the battel books listing his college accounts from 1760 through 1764 (when Mr Lyons was imported) and for a few weeks in 1765. Mr Hiscock, oddly enough, was a former Cape Horn sailor, having served an apprenticeship in the Scots four-masted barque *Inverness-shire* in the Californian and Australian trades. He was interested in Cook and Banks who, as Sir J. Banks, Bart., C.B., P.C., became well known not just from the *Endeavour* voyage and as an outstanding botanist, but as influential President of the Royal

* Mr Hiscock died at Oxford in 1966.

Society for over forty years until he died in 1820: but we could find no other trace of him – no painting, no plaque anywhere, no indication of his former rooms. The difficulty about these was the old system of sub-letting: Banks may not have occupied the rooms allotted him. Christ Church has welcomed many famous men: Banks was just one of them.

The surprise of the elderly professor at finding a gentleman undergraduate who actually *wanted* to learn something is understandable, perhaps, but it seems that Banks was not alone. It is reported that sixty undergraduates turned out for the first botany lecture which would be astonishing even today, though perhaps they were aware that this was history. Or they admired, or were intrigued by, the forthright spirit of the otherwise unscholarly Banks.

Banks had other sources of influence besides his own commendable acumen. A cousin, a beautiful girl, had married a Grenville: their daughter became second wife of the third Earl Stanhope. Another charming relative was wife of the eighth Earl of Exeter. Grenvilles, Stanhopes, Cecils had been of England's great for centuries, and still are. Though not a relative by marriage or anything else, the fourth Earl of Sandwich, John Montagu, was both London and East Anglian neighbour. They shared the interests of botany and fishing and were firm friends. The Earl had been a Lord Commissioner of Admiralty at twenty-five and First Lord twice prior to the *Endeavour* voyage, and again for eleven years afterwards. He was also a principal secretary of state for three years from 1762.

This was the kind of influence that counted. A note to Sandwich from the Royal Society requesting permission for its member Joseph Banks (very new then and still to make any contribution to knowledge) to accompany the planned expedition to the South Sea, with his suite of seven persons, was sufficient to crowd the *Endeavour*'s great and sundry other cabins. That they must also add considerably to the strains of the voyage on their captain probably occurred to none of them.

On 21 July 1768 His Majesty's bark *Endeavour* warped out of Deptford basin (where nothing whatever recognizable to

The Endeavour *on her way*

Cook remains today), slipped down to an anchorage in Galleons Reach to take in guns and ammunition, passed on through the shipping to the Downs and so to Plymouth, taking over three weeks to get there.

Among the navigational equipment aboard a conspicuous absentee would have been most useful. The dead hand of Edmund Halley and the live hand of Nevil Maskelyne the current

Astronomer Royal, committed lunarites both, stretched out to keep the Harrison chronometer off the vessel, which could have profited from it greatly. Lunar problems were useful in working out longitude for astronomers: the Harrrison chronometer solved the matter for working seamen. Fortunately, Cook was both. He could do without the chronometer, but it was a pity. Though both Halley and Maskelyne had put a great deal of work into simplifying their problems, lunars were really little help to seamen. Seamen knew that the establishment of a ship's longitude would be a simple matter if there existed a perfect timepiece which could keep Greenwich time throughout a voyage. There was a 'Board of Longitude' set up in England in 1714, with £20,000 as reward for anyone who could establish longitude at sea, within thirty miles. By 1735 John Harrison – a Yorkshireman like Cook, the son of a carpenter: no relation to Halley's anathema, the first lieutenant of the *Paramour Pink* – had produced a useful chronometer, but the Board, being faced with a solution not foreseen by its principal 'experts', was not impressed. Harrison had no influence either: but he persisted. By 1761 his hand-made chronometers were so obviously revolutionary and good that Admiralty agreed to test one in a frigate on a voyage to Jamaica and back, during which it lost less than two minutes at a regular daily rate, and made it possible to establish the ship's longitude by a few simple, properly timed observations of the sun's altitude, within less than twenty miles. This was a few years before Cook's first voyage.

Fortunately for Cook (who soon made himself a considerable lunar expert anyway) he was spared one serious navigational problem. The plan was originally to set up an observatory for the transit of Venus somewhere in the Marquesas Islands. First he would have to rediscover these, which existed only as vague and incomplete shapes drawn on a large chart by Mendana well over a century earlier, and had not been seen since. While the *Endeavour* was fitting out, Wallis returned from the second *Dolphin* voyage. He had discovered Tahiti and had established at least its latitude accurately. Tahiti seemed a pleasant place and had a harbour. The Marquesas were forgotten, and the first objective changed to George III's new island. The *Dolphin* made other useful contributions, including several officers, petty-officers and seamen, and a handsome goat said to be quite a sea-'dog' and a good milker. This unusual goat had already survived one circumnavigation: now it was to make another. Many goats went to sea in those days aboard Indiamen and the like, but the *Dolphin*'s was the only circumnavigator.

At 2 o'clock in the afternoon of 26 August 1768 the *Endeavour* sailed from her anchorage in Plymouth Harbour. The ship had been at Plymouth since 14 August, while stores continued to pour aboard and shipwrights worked away to convert the 'gentlemen's cabbins'. Mr Banks and Dr Solander, having posted down late from London, were the last persons to join, but everything was ready for them. By this time the ship was almost bursting her stout wooden sides, with ninety-four persons aboard from the ages of sixteen to forty-eight, ten carriage-guns and twelve swivels with ammunition, provisions and stores for eighteen months, and the two greyhounds tethered by the windlass while the mariners scampered about the decks and in the rigging, setting the sails.

The goat from the *Dolphin*, perhaps realizing that she was off on another circumnavigation, bleated forlornly. Too experienced an old goat to get in the way on sailing-day, the *Endeavour*'s small decks bothered her for she did not know where to go. So she got under the shelter of the longboat and lay down. The tide was just on the turn and there was a nice little breeze off the land.

CHAPTER SIX

With Stuns'ls Set

In its earlier stages, the voyage now begun was familiar to the point almost of being commonplace, for the *Endeavour* must take the same general route, at first, as all sailing-ships bound by the southern way, with the help of the trade winds, towards the West Indies and North America. Though he picked up the helpful trades long before ships which set out from Northern Europe, Columbus used that way, and there is considerable doubt that he was the pioneer. The *Endeavour* required an offing from the land – an 'offing' meant simply to get away, to put as many miles between land and ship as possible to reduce the danger of being blown back again by adverse winds, and perhaps wrecked – and Plymouth was an excellent place to start, for a ship there was already more or less finished with the Channel. Then she must get across the Bay of Biscay, which could be nasty at times, and pass down the coast of Portugal with a glimpse of Finisterre to check her position, then gradually haul to the westwards off the land. The winds are often northerly down the Portuguese coast, though there are plenty of calms and variables, and in autumn and winter nasty Atlantic gales can blow.

Every day at sea after Portugal brought a ship nearer the trade winds. The idea was to pick these up as soon as possible and stay in them. The expression 'trade winds' means more a zone of winds, N.E. in the northern hemisphere, S.E. in the southern, but not necessarily always regular. They were often met about the latitude of Madeira or earlier: west-bound ships for the Indies and America then began to add more and more westing to their course, to have the trade blow them – they hoped – right across the Atlantic, where the Gulf Stream would help to take them to North American destinations, or its avoidance help successful landfall in the Indies. If bound for the Cape

With stuns'ls set

of Good Hope, India or anywhere else in the East, or to 'the Brazils' – which meant in the eighteenth century the east coast of South America in general – then the ship ran with the trade wind towards the equator, keeping on the African side, passing within sight of one or two of the Canary Islands and the Cape Verdes to check longitude.

The north-east trade winds did not, unfortunately, usually blow the whole way to the 'Line'; sooner or later they petered out, giving way to an area of calms, catspaws, rain-storms and sudden, short-lived squalls, which seamen knew as the 'doldrums'. A square-rigged ship had to work herself through these the best way she could, by alert watch-keeping on her Master's and officers' parts in order to make use of every bit of wind whether favourable or not, and willing work by her seamen, who had constantly to be trimming yards and sails night and day, wet or dry. A ship could sit there looking at herself reflected in dead calm for days and even weeks, as if all wind or hope of motion in the air was gone from the world for ever. This was trying. Knowledge slowly grew (for those who sought it) of the best approximate longitudes to get across with minimum calm at various times of the year. When at last – perhaps in days, perhaps weeks – the south-east trades were found, all the square-riggers, by the limitations of their rig, had to do the same thing – take the trade on their port sides, keep their sails trimmed so that the ship sailed in a general southerly direction as close to the eye of the wind as possible (which with them was, at best, about six compass-points – 65 degrees – away) and, always keeping all sail perfectly set, jam along with the best speed they could make. Here skilful seamanship and helmsmanship were essential to keep the sails close to the wind and yet nicely full and pulling.

Square-riggers were sometimes known as wind-'jammers', and such close sailing was known as going 'by the wind', or 'full and by' if the ship was kept going a fraction off the wind. There was plenty of both on most voyages, but the trades were usually constant and offered a pleasant break from harsher latitudes. Indiamen sailed through both trades like any other long-voyage ships: at the end of the south-east trades, they worked through the variables and made the best of their way towards the Cape, usually finding plenty of strong westerlies if they dared to go a little south. As for the *Endeavour*, Cook meant to touch at Rio: after that he would have to work his way down the coast of South America. His real problems would begin – or intensify – as he approached Cape Horn: there and afterwards he was on his own.

The little cat met a mixture of weather the first week, winds favourable and foul, seas rough and quiet. Banks was seasick and so were others, for the motion of the deeply laden ship could be violent if the wind and sea were against her. Steep seas slapped her plump bows and she bucked over them. No considerations of comfort had entered into Cook's reasoning in selecting such a ship: she was slow – maximum speed perhaps eight knots, generally more like seven – and her upper decks were cheerless and exposed, for they lacked even the slight protection of the usual high bulwarks. There was no wind-break at all. The rail was open to the wind and the sea. Older ships had offered at least a little protection for helmsmen and watchkeepers by means of their high quarters aft which sheltered the helm: the *Endeavour* was flush-decked and open-railed, and in a blow the sea smashed at her and sometimes over her as if she were a half-tide rock. She had not gone far before a Biscayan blow did a little damage and 'washed over board a small boat belonging to the Boat-swain and Drowned between 3 and 4 doz'n of our Poultry which was worst of all', says Cook in his Journal on 1 September, when the ship had been a week at sea. (An officer's Journal was not the ship's log, the keeping of which was the Master's responsibility. The log recorded what was done in matters of seamanship – reefing, setting, furling sail, sending masts and yards up and down, going about, heaving-to and all those things – and navigation, which covered distances off and bearings from known land sighted – if any – soundings, leeway, estimated speed made good and measurement by heaving the log, observations if any and noon positions of the ship if known. It also listed the broaching of barrels etc. of the ship's stores, especially rum. Journals were ordered to be kept by all warrant and petty as well as commissioned officers and midshipmen, and were supposed to give an intelligible account of the voyage, particularly its 'remarkable occurrences'. Underground passages beneath the National Maritime Museum at Greenwich, near London, are still stuffed with officers' journals from the sailing navy, by the thousand.)

The loss of those poor bedraggled fowls, flung so unceremoniously into the Atlantic before they had a chance to find their sea-legs, was a serious matter because they were aboard to provide eggs for the great cabin and the sick: it was only after being long past egg-laying that they found their ways to the pot. The ship's cats, Mr Banks's dogs, and the wise old goat berthed themselves in a dry corner of the lower deck, out of the way, and survived. Dry corners, indeed, were scarce: for the upper works leaked and required more caulking, which could not be done at sea. But the goat soon learned her way about.

The *Endeavour* was off Finisterre eight days out from Plymouth, which was not bad. There were porpoises which played round the ship to the delight of Banks and his team, who even in their sea-sickness began to study the natural history of the sea. In light winds and calms they dredged and netted all they could from overside: in calms, they were pulled round in boats, noting and preserving all the life their sharp eyes could see – plankton, jelly-fish, minutiae of all descriptions, and birds which sometimes were blown out by off-shore winds to perch for a while exhausted in the rigging and then to die. As the sun shone and the quarters dried out, the green baize cloth was spread on the great cabin floor and the naturalists took over, Buchan and Parkinson drawing away industriously wherever they could find a place to work (the great cabin's five large windows brought in an excellent light), Banks and Solander examining specimens, discussing, classifying the classifiable or giving names to the new, while Spöring wrote it all down. The servants kept the cabins neat and clean, shined boots, scrubbed cabin decks with beach sand (from Devon) and sea water, trimmed candles, helped the cook prepare dishes for the captain's table and in the pantry, searched the well-stowed hold for sundry items called for by their masters and invariably at the bottom of everything.

As the weather permitted, they moved up on deck but it was even more difficult to find room to work there than below. When she was the *Earl of Pembroke* the *Endeavour* managed very well with a handful of apprentices and four able seamen. Now she had forty seamen and a dozen marines as well, to say nothing of an afterguard of seven including the two watch-keeping lieutenants, Master, gunner, surgeon, and their assorted mates. There were also the tradesmen – carpenter, armourer, sailmaker, boatswain, clerk. The surgeon was qualified in those days by the Court of Examiners of the Company of Barber-surgeons, and was not classed as an 'officer and gentleman', had neither uniform nor quarter-deck rank, and counted with the other tradesmen. He was regarded as a craftsman, an amputator, remover of limbs, a wielder of crude instruments with such strength and skill as he could muster, and if he could operate on flesh and bone as well as the carpenter could on timber the ship's company was fortunate. They were much more fortunate if he was not called upon to operate at all. As his surgery was of use mainly in and after battle, he was not very busy with Captain Cook, though he acted also as physician (as well as he could) and apothecary. Eighteenth-century seamen were apt to regard him with suspicion in all three branches of his calling, and kept away if possible. In other ships his principal employer could be trusted to make an early appearance aboard and remain there. This was scurvy, then the great shipboard killer. Surgeon Monkhouse of the *Endeavour* – in fact an excellent man, and as competent as ship's surgeons came – was allocated a mate to help in fighting this, and in general. The ship also carried probably the most active and determined enemy of scurvy and all dietary ills any ship had known, in the person of Cook himself. She also had probably the most extensive array of hopefully assembled anti-scorbutics, some to prove more useful than others but all (and more) to be forced down the reluctant seamen's rum-sodden throats by the indefatigable and determined Cook. If there were any deepwater seamen who were *not* going to die of scurvy they were the crew of the *Endeavour*.

Surgeon Monkhouse oiled and greased his saws, amputating

knives, bone-nippers and turn-screws, rugins (files), tenaculums (a sort of hook for holding back sundry pieces of limbs which might otherwise interfere with amputations), forceps, needles etc., etc., and put them carefully away in the large chest provided (and certified) for their stowage by the Company of Barber-Surgeons, chocked the chest in his minute cabin together with his chest of drugs, began his Journal of Practice, listed the various remedies and comforts put in the ship by the Sick and Hurt Board (this took some time) and prepared his file of Sick Tickets, certificates without which no one aboard could be officially sick at all. Indeed the paper side of his work kept him busier than the practical (for this apportionment of Services labour is no new idea), and he was eager to get on with it, to have more time to spend with the interesting gentlemen, Mr Banks and Dr Solander. It was a privilege to have shipmates like these, and Surgeon Monkhouse looked forward to the enjoyment of their company and their knowledge.

The surgeon's mate was carefully checking the distilling apparatus (stowed then in the hold along with so much else), which was the same type that Captain Wallis had carried. Dr James Lind had developed a simple contrivance consisting of a large copper in which sea water was boiled and the steam condensed in pipes (which could be easily converted from old musket barrels), from which it was caught in bottles. Wallis's surgeon declared that he could make forty-two gallons of fresh water an hour from fifty-six gallons of sea water in a watch, by means of this apparatus. If so, he did extraordinarily well. There were problems, of course. On long voyages, perhaps the chief was fuel. This had to be mostly wood, and it could not be carried in quantities sufficient to do all the cooking and condense fresh water as well. However, fresh water was one of the most serious shipboard problems at the time, for neither water nor the contemporary beer could be made to keep in casks. Hence the excessive use of spirits – half-a-pint a man a day, sometimes with (but usually instead of) a *gallon* of beer – and sundry concoctions based on them such as 'flip', the preparation of which passed Gunner Stephen Forwood's time very pleasantly and their consumption, with cronies, more so. The Gunner wasted no time setting up a pleasant drinking gang which by pooling resources could be cheerfully drunk every evening. He probably regretted that the Purser was not also a member of this company, for then supplies might be even better. But Captain Cook was his own Purser: there was no nonsense there.

In such a ship the Gunner had not much to do save to keep the powder dry and secure, the guns serviceable, and the inevitable book-work more or less up-to-date. The *Endeavour*'s ten carriage-guns and twelve swivels (four for the long-boat, the others to be mounted on deck as and when necessary) were normally kept below on passage, though she was pierced with ports for the larger. These ports were tightly secured and caulked until she reached the Pacific, for the lower were close to the water-line and something of a menace. The Gunner, however, was an important man whose equipment could be vital: drunk or sober, his work was never neglected.

The Armourer had a forge and blacksmith's tools and was more blacksmith than anything else. There was plenty of ironwork to look after on the masts and yards: if there was not much else for him to do, he could always make those valuable items of trade in the South Sea – nails. John Satterley the Carpenter – 'an excellent man' – had plenty of work always, for he was also head cooper, head caulker, joiner, shipwright, and cabinet-maker. Cooperage alone was a full-time job, for the ship carried barrels, casks, hogsheads and puncheons of all descriptions, many of them in staves. They were vital and they had to be looked after. When the ship took on fresh water, it had to be by barrels brought out in boats. The Sailmaker's duties were obvious and his chief responsibilities in the constant sight of everyone. The *Endeavour* sailed with two suits of new sails as well as the old Whitby suit, all of flax canvas, hand-sewn. As well as looking after the sails, the sailmaker was in charge of hatch and skylight coverings and all those things, and he cut and sewed canvas wind-chutes called wind-sails, cylindrical ventilators open and winged-out above decks to funnel fresh air below (to the annoyance of the mariners who thought such draughty stuff should stay above deck where they were used to it: a man should sleep in a fug) and help to keep the lower deck sweet. Cook was very keen on ventilation, and so were his excellent officers Hicks, Gore, and – later – the promoted Clarke. Lieutenant Clarke had served as A.B. and midshipman with Byron in the *Dolphin*, and Lieutenant Gore was Master's Mate in that frigate on both the Byron and Wallis circumnagivations. Lieutenant Hicks was a consumptive, and he found his hutch of a cabin deep below a considerable trial. It probably helped to kill him, for he died long before the voyage was over.

These veterans, and several seamen who also came out of the *Dolphin*, were a great help to Captain Cook, who had copies of the Wallis and Byron journals aboard from which he learned more, perhaps, than their writers intended.

So the little ship settled down, and others of the score-odd craftsmen and tradesmen (including the cook, two quartermasters, and three midshipmen whose duties were defined but not always performed), the eight servants and the forty able seamen, established their routines, their social groups and shared pastimes and communities of interest which would help to fill their long days and add meaning to the voyage. It is difficult for us nowadays to appreciate how so many found so much to do in so small a ship, but seamen have always been adept at organizing their crowded lives. Just to give each other room to live and to keep sufficiently out of each other's way was an accomplishment calling for considerable experience and effort. The official allowance of slinging space for the mariners was fourteen inches, so they had to sleep on their sides. Above decks, there was always the rigging to care for – an endless job which was particularly arduous in newly rigged ships. No matter how well the fitting-out riggers might have done their work, invariably there was a great deal of chafe to put right, as well as leads of gear to improve, blocks and tackles to free or to make run better, and the set and trim of sails to be continually revised.

The complicated array of a sailing-ship's rigging, slowly evolved by painful trial and often fatal error down the centuries, was not then perfect and never became so while square-rigged ships lasted. As topmasts with topsails were added to the original lower-masts and then topgallant-masts with more sails stepped at the topmast-heads, and masts increased in number from the original one to two and three (and even four), bowsprits lengthened and optimisitc innovators (a rare breed then) thought of adding studding-sails set on extensions rigged out precariously on the already burdened yards, very obviously the wind-using ships became more complicated. The stresses of everything had to be carried upon a constantly moving base. The essential rigging to support the masts (called 'standing' rigging because it stood where it was set up, to support the masts) and the pattern of running rigging (it 'ran' through blocks, bull's-eyes, or thimbles, at its work of hoisting and trimming yards, setting sails, 'clewing', and 'bunting' them up again) had grown with the masts, the essential gear to manage new sails being contrived, rigged, and named with the sails. All things followed ancient patterns and a nomenclature precise, voluminous, and grown almost sacred by long usage, the very essence of the sea. A seaman from the *Endeavour* could have served in Drake's *Golden Hind* with no difficulty, though a twentieth-century steel ship man might take a few hours to get his mind to accept so complicated a simplicity – complicated because of the profusion of pieces of hemp to do work better done by a little wire or a few more efficient tackles, but essentially simple because the work it all did was so obvious. There was a precision, an exact terminology, in use throughout the sailing-ship era which once learned stayed with seamen. After all, their safety and lives often depended on the instant and unquestioned understanding of a few familiar words, in set order, flung at them above the tumult of wind and sea.

So it was with Cook, himself high among the supreme practitioners of the science and art of sailing-ship handling. With so much that had to be done by good men who knew their work, there was no need for drills. Aloft all day, the rigging was filled with brawny mariners like migrant birds resting on some familiar tree: but they were not resting. Their thick tarry fingers got skilfully on with work there, rough and intricate. Their mellow shouts, bawled strictly as necessary, floated to the deck. Loose this! Leggo that! Heave on the other! Stand from under! Bela-a-ay! The boatswain and the Master's Mates kept an eye on them, and so did the lieutenant of the watch, while looking out endlessly for vagaries of the wind *before they reached the ship*, to add his own sea lingo with the essential orders in good time. The square-rigged ship, with her hostages to fortune spread high aloft, had two great problems – first, to keep the sails and rigging there and not to be dismasted: the second, to make best use of every breeze that blew. The helmsman was busy at the wheel before the mizzen-mast, watching his compass course and, when he could see it past the sails, the swing of the bowsprit-end against the clouds, while controlling the rudder's angle through a cumbersome set-up of hempen ropes led from wheel-drum to tiller-head. He, too, knew that his duty was to correct for inessential meanderings and other diversions of course preferably before they could really get started. Awkward as the tiller ropes were, yet he had the ship's 'feel' in his hands, and carefully watched his work. One of the quartermasters watched him, for good course-keeping was vital where dead-reckoning was often the only means of estimating position. Lunars were still irregular and difficult: and the Harrison chronometer was left further astern every day. By the taffrail aft the midshipman of the watch, with two young able seamen, was 'heaving the log', called that because they were actually 'heaving' a 'log' (piece of wood) over the side into the sea – not to throw it away, but to measure the ship's speed. A triangular piece of wood was secured to the end of a light piece of good hemp like signal halliards, which was marked with a knot every two-and-a-half feet or so. The method of use was for the Midshipman to hold a half-minute sand glass which he turned as the first knot, pulled from the coil held by one of the seamen, entered the water. At the precise second when the sand all ran out the Midshipman cried 'Stop!' The running line was checked by the second A.B., the number of knots which had entered the water counted, and that was the ship's speed in 'knots' as at that moment. The knots on the line were each carefully positioned to indicate the passage of the ship through one sea mile of water: for the slow speeds of eighteenth-century sailing-ships the method was accurate enough. Like so much aboard, the manner of its use and its purpose were immediately obvious: and it was cheap and easy to maintain.*

Arranging better stowage below, better leads aloft, maximum organization of the decks and maintenance of scrupulous cleanliness everywhere while the trusted tradesmen each looked after their own departments, and the interesting supernumeraries aft amused themselves to the increase of knowledge happily gained, the little vessel lumbered along her tremendous way. The crew was a good crew. Cook was not senior enough to be allowed to choose his own men, but five came with him from the schooner *Grenville*, including his personal servant. The obvious attachment and respect of this quintet for their dour captain helped to establish his reputation aboard. Cook was a hard man to know and could be very short-tempered over slovenly seamanship in any guise: but the lower deck soon knows its officers.

Of those of the crew whose home towns are recorded, there were seven Scots, several Irish and Welsh seamen, as well as a Venetian, a Brazilian, and a midshipman from New York, named Magra (sometimes written as Matra or Marra, for the eighteenth century was careless of its spelling). Mr Magra etc. was rather an unusual midshipman, of whom we shall hear more. The crew, as well as the officers, included several veterans of both the Wallis and Byron voyages, which is perhaps odd, for they did not have to go. They were career men aboard voluntarily, and they must have taken a liking to exploration. Wallis looked after his men and so did Byron, for they were both humane officers and pleasant men. It was as discoverers that they might have done better. The able seaman did not get much out of existence in those days beyond the satisfaction of the sailing-ship life on a worth-while voyage under good officers: all those forty A.B.s were professional seamen, properly qualified (most of them) in the merchant service as well as the Navy. They were obviously also adventurous. Of course, a few to whom the voyage offered no attractions deserted while the

* This method of measuring speed was still in use in big sailing-ships in the 1920s.

ship lay at Deptford, but they were not missed. Whoever deserted Cook's *Endeavour* was well lost.

The servants too, were unusual young fellows, like twelve-year-old Isaac Manley, son of a bencher of the Middle Temple, signed as Master's servant, who was a midshipman in the *Resolution* on Cook's second voyage while still in his very early 'teens, became admiral, and an honorary D.C.L. of Oxford before he died in 1837. Another little fifteen-year-old, who already was a veteran of the *Grenville*, also made admiral. This was Mrs Cook's cousin Isaac Smith, and he was the only man that nepotism helped to put aboard. The oldest member of the crew was the Sailmaker, a Yorkshireman named John Ravenhill. His age is listed as forty-nine but Cook and his shipmates thought him very much older – perhaps eighty. Old sea-dogs before the mast were rare then. Life for them was short and usually made passably merry on a brief and very temporary basis only by a large intake of 'flip', preferably on a nightly basis. Old 'Sails' kept himself well soaked, apparently, but he knew his work and did it well. His whiskered, gnarled old countenance and his bleary eyes following the stitch-stitch-stitch of the sail needle as he sat bent on his bench, a sail before him, were as much a part of the deck by day as the oddly carved face on the windlass. Smoke from the cook's galley, back-winded on him by a flap of the fores'l, caused him to use good sea lingo of the early eighteenth century: but he was a quiet man, even when full of flip. He was to stitch many of his shipmates into their canvas shrouds after the fevers of Batavia caused fatal illness: he was the only man aboard not to be infected there. Perhaps the germs could not fight their way into his pickled old carcass. He, too, died before the end of the voyage, but it was of old age.

All these tradesmen were key men, but perhaps the most important was the cook, one John Thompson, another Scot probably, age unknown, but a much-scarred veteran of sea battles when he came aboard at Deptford one day in June. Among other things, he had lost his right hand. This was considered little if any handicap to a naval cook, for the curious

English contempt for the entire maritime food industry and all in it was then very strong in sailing-ships. One reason for this, in the Navy, was the unfortunate dishonesty of pursers, contractors, purveyors of ships' provisions and all such persons. There was a strong tradition that – maybe on the score that seamen died young anyway and therefore what they ate was of little consequence – provisioning was a God-given means for these devil's disciples to line their pockets, which they did. Salt beef in the sailing-ship era was known as salt horse with good reason, for horses' hooves and shoes were sometimes found in the casks. The seamen did well if the 'beef' was nothing worse than horse-meat. The whole seafaring food industry was almost thoroughly ghastly.

It followed that, since it mattered very little who prepared the stuff, anybody could be cook – that is, anybody not fit for much else. Instructions drawn up for sea cooks in general* in the mid-eighteenth century are brief. Here they are, verbatim and in toto.

Of the Cook

1. He is to take upon him the Care of the Meat in the Steeping-tub.
2. In stormy Weather, he is to preserve it from being lost.
3. He is to boil the Provisions, and deliver them out to the Men.

That is the lot – no baking, no frying, braising, grilling or anything like that, not even a smell of fish and chips – indeed, no chips at all. The French baked bread at sea and baked it well. The Spaniards, the Portuguese, and the Italians made a hundred dishes out of preserved sardines, anchovies and the dried cod called *bacalhau*, with rice, powdered tomatoes, olive oil, and garlic (which never decayed) when fresh food gave out. The Hollanders had a hundred seagoing fish dishes too, with North Sea herring salted, dried, smoked, pickled or just raw in strong vinegar, and eels, to say nothing of twenty kinds of cheeses and all sorts of palatable sausage. But English in-

* Printed in the *Sailor's Companion and Merchantman's* Convoy . . . *Shewing the Duty and Conduct of all Superior and Inferior Officers of the Royal Navy of Great Britain*, etc., by one J. Cowley (London, 1740, printed for T. Cooper at the Globe in Paternoster Row).

genuity seemed to give up in this matter of food for mariners. Instead of bread there was 'biscuit' – hard-baked stuff put aboard in sacks in which weevils prospered and multiplied – and instead of staples there were salt 'horse' and salt pork. The *Endeavour* had a fish room built in the lower hold at Deptford, but what went into this is not recorded. The English palate never took kindly to boiled *bacalhau* or raw herring, but there must have been some salt cod aboard. For the afterguard there were luxuries like the occasional fresh egg, a little spiced meat, 'orange conserve' in stone jars (marmalade) and north-country cheeses likewise, sauces (well developed, perhaps because the food was so indifferent), dried fruits and raisins and 'Portuguese figges'.

'A sea-cook,' wrote that scurrilous maritime historian Ned Ward in his *Wooden World Dissected*, etc.* in the 1750s, 'cooks by the Hour Glass as the Parsons preach Sermons. ... His Knowledge extends not to half a Dozen Dishes; but he's so pretty a Fellow at what he undertakes, that the bare Sight of his Cookery gives you a Belly-full. ... All his Science is contained within the Cover of a Sea-Kettle. ... He's an excellent Messmate for a Bear, being the only other two-legged Brute that lives by his own Grease ...'

These were rough words, but they were justified. In larger ships there were captains and officers' cooks who could do better and, on short passages, live beasts were carried for fresh meat. The seamen, when they could, fished overside industriously. But oxen could get scurvy too, or at any rate thin down to uselessness without it, and sheep took poorly to the sea life. This left hens, hogs, and goats: in good weather the poultry prospered. The goats prospered always.

Captain Cook gave his meticulous attention to the provisions as to everything else. The cooks sent by the Navy Board were the only crew to whom he took objection. The first had one leg. Cook objected to such a 'lame and infirm' horse-boiler, the more urgently 'as he doth not seem to like his appointment' and begged the Board 'to appoint a nother'. The Board obliged, but the 'nother' was the left-handed Thompson. Cook again

* Edinburgh, printed for James Reid Bookseller in Leith, 1752.

objected, thinking that all the hands should at least be whole: but the Board, having 'no Ship Vacant to provide for John Thompson Cook of the *Endeavour* Bark', ordered that he must stay. 'Nother' Thompson himself made no objection to his appointment. Under a commander who took an interest in him and his department, and a surgeon who shared the same unusual keenness, he really tried, and did very well, in spite of the fact that he had to handle all sorts of new-fangled stuff to keep down the scurvy: Sour 'Krout' (pickled cabbage) by the dozen barrels, malt in hogsheads, Rob of Oranges and Lemons (a sort of concentrated juice), Portable Soup (solid blocks of extract of meat, very nourishing), Wort (the infusion of malt before being fermented into beer) and Saloops or saloups (for making hot drinks from sassafras, which used to be sold at London's street coffee-stalls as being cheaper than coffee and alleged to 'do good') were put aboard by the Sick and Hurt Board, and the cook was instructed how to deal with some of them. Mustard seed and vinegar came from normal stores.

Trying to do something really useful about scurvy in the Navy was no new idea. The famous Scot, Dr James Lind, 'father of nautical medicine', had done a great deal of work and research on the problem and could have driven it out of the Navy had he been allowed: ineptitude, disbelief, maladministration ashore and afloat, stupid acceptance of time-honoured ills which had become part of the sea life, and lethargy of responsible committees, apathy of seamen and indifference of officers – these things stopped him. Cook's friend Palliser had made long voyages without losing a man to the unnecessary disease, years earlier. So had a few others – but very few. As one of his seniors pointed out, James Cook was not only thorough, but came from the 'class' which was accustomed to obey orders: therefore the policy of loading his 'bark' with all this experimental stuff was justified, for it would be properly tried out and reported on.

Some of these preventives were better than others. The 'portable soup' – so called because the slabs of it were readily portable – was probably the best. The Sick and Hurt Board sent a thousand pounds of the stuff and laid down that it was to be served on the three meatless days each week 'to the well men as well as to the sick'. These meatless days were Mondays, Wednesdays, and Fridays (after 'bully-beef' in tins was developed, these were reduced to two). The mariners got their new soup boiled for fifteen minutes with either pease or oatmeal in the proportions of an ounce of soup, a quart of water, and four ounces of pease or two ounces of oatmeal for each man. It proved 'extreamly beneficial'. The weekly allowance of two pounds of 'Krout' for each man went down well. This was served with the salt meats. At first it did not appeal to the mariners at all. They spat it out and swore. But Cook had been a mariner, too. He simply added more 'krout' to the cabin table where the gentlemen, under instructions, ate it with such relish that reports trickled for'ard that the stuff might be edible after all. After a week or so of this, the cook reported that the mariners were requesting their share of Krout. After another day or two, they were allowed some. It went down very well then for the rest of the voyage.

There was a great deal more to the health of his men besides diet, as Cook well knew. Rigorous and continual attention to cleanliness and ventilation below were essential, and his officers and all the makers of rounds – inspections – were at once instructed to see that standards here were the highest possible. He was determined that they would not have to make their rounds below decks with a silver spoon held up to see how much the foul air tarnished it, as was done in many of the big ships in bad weather, when their gunports were secured and hatches battened down. When the spoon tarnished too badly it was taken as indicating risk of suffocation and a gunport might be opened. Properly maintained general health helped to prevent scurvy. The mariners had to wash in sea water and keep themselves clean as well as they could There was plenty of sea water, reasonably warm after the first week. Bedding was aired regularly, utensils kept scrupulously and the cook's coppers burnished to the standard of the captain's speaking trumpet.

Some of the mariners murmured, of course: but the *Grenville* men told them they had better put up with it. There was more than a murmur about innovations enforced in the galley, where it had long been the practice for the cook's cronies to receive favours, such as some saved beef fat from the salt-horse days which they worked into their personal puddings or rubbed into their biscuits, warmed at the cook's fire. The biscuit was dry and unpalatable stuff: each man had an allowance of a pound of it daily, and munching this became monotonous. Saved-up ship's beef-fat, however, was a scurvy-breeder of considerable merit: its use in food was therefore forbidden. Whenever he could, Cook introduced fresh foods of all or any kinds, from stuff called wild 'Sellery' to odd-tasting 'scurvy grass'. One of the troubles about a diet of salt horse is that it quite ruins the taste; the mariners grew in time to like the stuff they became accustomed to and – worse – abhor all else. Their abhorrence of 'sellery' and scurvy grass was in line with their dislike of all innovations.

With Cook they learned to eat the ghastly stuff or take the consequences. At Madeira, Able Seaman Henry Stephens, from Falmouth, aged twenty-eight, and Thos. Dunster, private of Marines, had the temerity to refuse their doubtless somewhat tough fresh rations of beef, in an insubordinate and noisy manner likely to infect others. Duly found guilty, they were paraded on deck, stripped to the waist, secured to an upturned grating to receive a dozen of the best, laid on with might by the bos'n's mate while the other eight marines were drawn up under arms in awesome square, under their sergeant John Edgcumbe, and Drummer Rossiter beat solemn time to the lashes. The lesson was salutary. Refusal of fresh rations after that was done very quietly, if at all. As for the lashes, these were standard punishment of those days and for almost a century afterwards. Most other commanders – if they noticed the offence – would have given at least two dozen. This was the eighteenth century, and no one then thought with twentieth-century minds.

At Madeira an otherwise pleasant stay of six days was made, while the ship took in something over 3,000 gallons of fine Madeira wines – chief reason for the call: all naval ships dropped in for Madeira wine – and all the fruit that would keep, as well as thirty pounds of good onions for each man. The crew also got all the sails up from the locker and dried them in the warm sunshine, caulked the sides, filled the casks with fresh water, and hoisted up a beef steer of 600 pounds in weight by the longboat tackles, but the three dozen fowls washed over the side were hard to replace. As for the steer, it was slaughtered before leaving and served for several beef days, well boiled.

While working anchors off Funchal – a deepwater roadstead where small ships had then to moor well inshore – Mr Weir, Master's Mate, was caught in a buoy-rope while leaning outboard working, carried down with an anchor, and drowned. There were no seamen offering ashore, so Cook exercised his right to impress from merchantmen – steal from their crews – and sent a lieutenant and party to seize one Jno. Thurman, Seaman, from a sloop belonging to New York. This was a British-flag ship at the time: a sloop was a small full-rigger. What Jno. Thurman thought of his fate no one bothered to record. He must have been a good man as he was soon selected to be sailmaker's mate. The name of the sloop he was impressed from is not now known: she had to put up with it.

After Madeira, wandering down through the north-east trades with a sight of the peak of Tenerife and Boa Vista in the Cape Verdes, Cook showed his humanity and his forward thinking by putting the mariners on three watches, a sensible step which commercial sailing-ships, not having large crews, were never to follow. The normal two-watch system meant that no seaman could ever have more than four hours' rest while the ship was at sea, for the watches changed punctually and completely at the running-out of eight half-hour sand-glasses. Half the seamen were on deck all the time. While this was necessary or at least advisable in bad weather zones, the *Endeavour* was an easy ship to handle alow and aloft and there were men enough aboard to strike her down to lower-masts at every eight bells, and rig her completely again in half-an-hour after. In good weather zones this was ridiculous. Cook therefore divided the forty A.B.s into three watches of twelve each, gave the other four various day jobs, and let his men have eight hours in their hammocks by night. The simple, welcome rest of a privilege such as this made a considerable difference to health and morale, for the men were properly rested for probably the first time in their sea lives.

They were issued with 'pipes and tobacko and fishing geer for everybody' for they had the odd idea that hot smoke drawn from clay pipes did their hairy chests good, and the fishing 'geer' might add to fresh food supplies. As the ship wandered along on her lonely way through the tropic seas, little fish learned to gather in the shade beneath the counter, and big fish often chased the flying-fish turned up by the roll of foam noisily spreading from her bows. Sometimes by night flying-fish flew in through the open stern windows, to the delight of Banks and Solander.

This was glorious sailing. Nobody noticed any longer that the old cat was perhaps a little slow. The sun shone and the shadows of the sails spread upon the foam of the ship's passing, as she rolled gently and let her timbers creak a little in easy acceptance of the sea's stresses. Draughtsmen-artists Parkinson and Buchan now had their easels up on deck all day long. Naturalists Banks and Solander observed and noted everything with unalloyed delight. He hadn't foreseen, said Banks, that the voyage would give him so splendid a chance also to study the natural life of the sea, as well as islands. He and Solander and all their party were busy shipmates, pleasantly occupied, and the atmosphere in the great cabin was as happy and warm as the sunshine streaming in the stern windows to light the green baize cloth on the well scrubbed deck. In the evenings candles were brought out for journal writing, while a decanter of Madeira passed around and the gentlemen conversed together on matters of natural science and other topics. Banks and Solander were the first non-service gentlemen, probably, and certainly the first university men that Cook had had the chance freely to mix with. He learned a lot from them. So did they from him. A week in the Doldrums bothered nobody and, once across, the passage of the south-east trades went by in a flash.

At Rio there was a bit of a shock. The *Dolphin* had been well received there, but the governor, a military gentleman with no knowledge of ships, could not believe the old North Sea collier was a king's ship and refused to allow the officers to go unescorted ashore. This was galling, particularly to Cook and Banks, but the governor was hardly to be blamed. There were then, and for years before and afterwards, many foreign smugglers and clandestine English traders on the coast of Brazil: his advisers, at first glance, classed the *Endeavour* with these or – worse – as a commercial spy. Further glances confirmed their suspicions. They pointed out that the ship had few guns and no proper gunports, and her hull was not that of a ship-of-war: her after-guard did not wear the king's uniform (the governor was not used to erudite globe-trotters turning up in naval ships, or any others): the crew was not drilled but got on with such merchant-ship tasks as shifting the topmasts, caulking the topsides, and overhauling the rigging in general (in preparation for Cape Horn): and their 'yarn' about sailing to the Pacific to 'observe the transit of Venus across the disk of the Sun' at some unheard-of island was regarded as a yarn indeed. Which, of course, essentially it was: prior rights of discovery and flag-raising in *Terra Australis* was the real objective. Banks managed to slip ashore from an open stern window into a boat by night, and did good botanical work, but Cook, for the most part, remained austerely aboard. Like a good naval officer and surveyor, he also managed to make an excellent surreptitious chart of the harbour. The mariners had to stay aboard, except for boat-work. The pressed man from the New York sloop turned surly for a while, perhaps because he had been in Rio before and knew what they were missing. He refused duty, which was mutiny. Again the grating was rigged, the marines fallen in and the drum rolled, while a smart dozen taught Jno. Thurman not to misbehave aboard an H.M. ship. Worse, one of the *Grenville* men – that 'good hardy seaman' Peter Flower who had been with Cook for five years – was drowned. (Few of those eighteenth-century mariners could swim: if their ship sank they liked to sink with her fast, if they had to.) There were cork life-jackets served out for the boats, but they were against them too, on the grounds that they were cumbersome and made boat-work difficult. A Brazilian seaman named Pereira was shipped in Flower's stead. Whatever the governor did, considerable supplies of beef, greens, and yams were bought daily and Cook managed to buy his fresh vegetables, meat, and fruit for sea stock. The ship had been a few days over ten weeks sailing to

Rio from the Channel. There was no sign of dietary deficiencies, but it would be a year or more before she could hope to be any where near a civilized port again.

It was 7 December before the *Endeavour* slipped quietly out of the troublesome port, made an offing, and stood down towards the island called by the Hollanders Staten Landt and now known as Staten Island, some 2,000 miles away. Her course was slightly west of south. This was a trying passage to make, though it was then the southern summer. The winds were variable at best, the current adverse, and sudden squalls could blow up viciously and without notice. No scudding cloud nor darkening horizon gave token of their presence: later sailors called them *pamperos*, and feared them, for they could blow all the sails out of a ship in an instant, or dismast her if the sails were too strong and she was too stoutly sparred.

Cook's passage in the *Endeavour* to the Horn and to wind'ard past that grim headland – the only wind-'ard passage he ever made there – was a classic. Those famous Cape Horn master mariners, the Germans Robert Hilgendorf and Hinrich Nissen, the Finn Ruben de Cloux, the Britishers Learmont and T. C. Fearon (who between them fought square-rigged ships past the Horn more than *two hundred* times) could not have done better in their great steel ships. Already when Cook went to do battle with the Horn the place had a bad name – so much so that the very difficult passage through the Straits of Magellan was preferred. This was plain bad seamanship, built on fear. It is a lubberly idea that such a strait offered a better means of sailing west than the open sea did, off the Horn. The trouble with the Horn route was that it was *too* open, for the ghastly succession of tremendous storms which raced round the world down there flung themselves upon the gaunt land which stretched its cold hand towards that home of the blizzard, Antarctica, as if they wanted to blow it over, and such a sea arose upon the banks by Diego Ramirez and all Tierra del Fuego as threatened to overwhelm the stoutest ship. *All* these great storms came from the west, right in the paths of ships trying to reach the Pacific from the Atlantic. The passage of these ships was often a long-continued, relentless fight against violent wind and appalling sea. A sailing-ship fights thrice as hard when she must head into the wind and feels gales then infinitely more, and her sails become crude aerofoils, like a giant aircraft's wings. Wind behind the sails, from abeam to right aft, gives a glorious shove that can be tolerated and managed even in hurricanes, so long as the sea runs true and is not flung about in mountainous, snarling heaps by twisting squalls. The ship flees from favouring winds. Their violence, though still awful, is minimized. But the wind ahead, striking the weather leeches of such sails as she might then bear, screams with the threat and the sound of hell. The 'lift' that gives her forward speed comes from the passage of the wind across her angled sails; her thrusting bow digs deep into the opposing seas which race at her in alarming, broken hills ready to smash over her. Only great steel Cape Horners like the powerful five-masters *Potosi* and *Preussen*, and their four-masted kind, could stand much of that over the years. The sailors in such ships – in all engineless square-rigged ships – reckoned their roundings of the Horn in westwards passages only, for the other way they regarded as down-hill, running with the wind. This could be difficult but by comparison was not worth counting.

All this was bad enough: but at least a ship off the Horn was in the open sea. She could yield, lie-to, shorten down until she shouldered the sea like an albatross asleep, hove-to, no longer fighting: she might drift or even drive to leeward, losing precious miles she had fought furiously to make. But she was safe. She would survive to come back and take up the fight again. Her people, save for a minimum watch on deck, could sleep a little while the fight was temporarily off.

In the Straits of Magellan there was little sea-room. There were far too few anchorages suited to the non-powered ships' primitive ground tackle. This was a tortuous place of twisting channels and wild 'williwaws' of wind leaping suddenly and malevolently down from the mountains to the danger of ships: there was often poor visibility through which ships groped at their peril. If they had no anchorage they had to keep under way. The wind was generally from the west, shrieking round the headlands and through the valleys. The hempen cables which

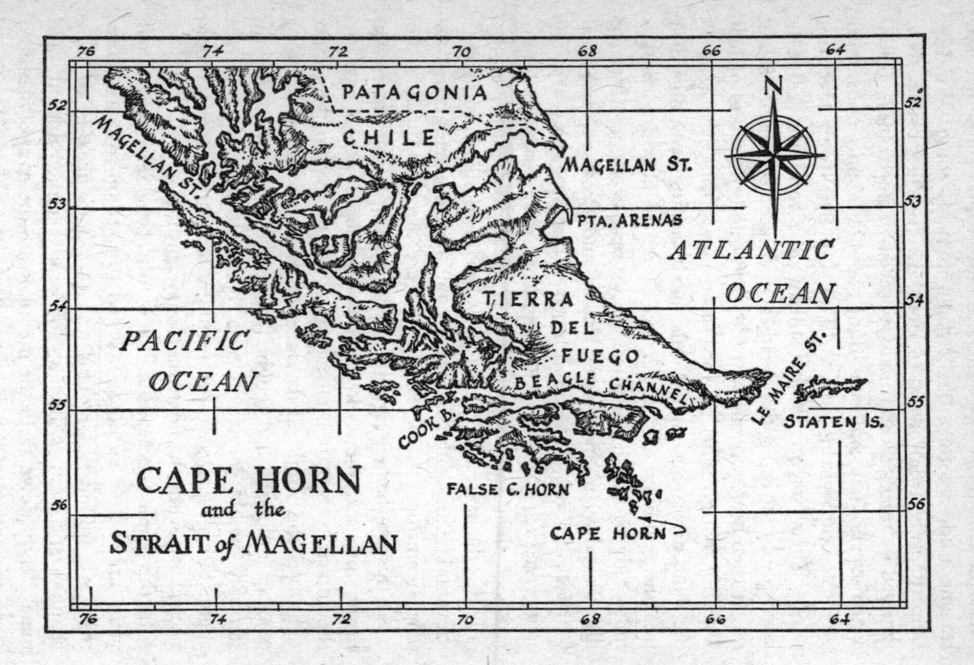
PATAGONIA
CHILE
MAGELLAN ST.
MAGELLAN ST.
PTA. ARENAS
ATLANTIC OCEAN
PACIFIC OCEAN
TIERRA DEL FUEGO
BEAGLE CHANNEL
LE MAIRE ST.
STATEN IS.
COOK B.
FALSE C. HORN
CAPE HORN
N
CAPE HORN
and the
STRAIT of MAGELLAN

held eighteenth-century ships to their big anchors were vulnerable on rocky bottoms, where they chafed; but often the ships had to anchor for days or weeks on end because they could not progress. The Straits were poorly charted then, too; there were many turnings and it was quite possible to take the wrong one. Arrived at last at the Pacific end, westerly gales still blew there far too often, making it difficult and dangerous to beat out and all too easy to be blown back again. Byron and Wallis both took months to get the fast-sailing *Dolphin* through, though Wallis suffered some delay because of the poor sailing of Carteret's *Swallow*.

The Straits had some advantages, of course. Generally the surface was much flatter inside and there was an entire absence of the enormous seas of the Horn. A ship *could* anchor now and again, too, and rest herself and crew, without losing ground. Here and there ashore a bit of firewood offered, or some wild 'sellery' and such stuff: but the wretched Patagonians who were occasionally seen were on too low a subsistence diet themselves to exchange any food.

Cook went the way the Californian clippers learned was best and Hilgendorf and all the others were to use – through the Straits of le Maire between Tierra del Fuego and Staten Island, and then making all the westing he could as quickly as he could. The Horn westwards passage meant a sail over at least 1,500 miles, from 50° S. latitude in the Atlantic to 50° S. in the Pacific, for there was a horrible lot of westing to be made before a ship was clear. 'Cape Horn' meant to seamen not just one clear-cut corner but a long-continued extremity of the South American continent and its labyrinth of off-shore islands. The Horn itself was the southern tip of the southernmost island, beyond which was the wide passage known as Drake Strait (called that because Drake, blown away south after passing through the Straits of Magellan, surmised that open sea was there). Shore-based advice offered at the time was to pass to the east of Staten Island and then make southing as far as 60°. Why make south when it's westing that's wanted? asked Cook, very sensibly. Why avoid the Straits of le Maire, which were short, when one tide would see you through? There were anchorages, too, for any who dared to use them. He was two months out from Rio before he was in the Straits of le Maire. For the first time on the voyage, the *Endeavour* had to heave-to in an offshore gale which blew 'vastly strong'. Banks noted that 'the seamen in general say they never knew a ship to lay too as well as this does, so lively & at the same time so easy'. He noted also that on Christmas Day almost all hands were 'abominably drunk', which was in keeping with the times. It was the only way they knew to celebrate. The rigging had been set up, new tiller ropes rove, the few carriage guns on deck unrigged and sent below, and the mariners each issued with a 'Fearnought Jacket and a pair of Trowsers', after which they did not complain of the cold. The galley fire when lit kept the fore part of their quarters warm: for the rest, there was no heat at all except for a small stove rigged temporarily in the great cabin. It was midsummer, but the snow lay thick on the Patagonian hills and all Tierra del Fuego. When Banks took a party ashore from an anchorage near the most easterly point of Tierra del Fuego – the sort of place only Cook would anchor at and only Banks land – they were benighted, and the two coloured servants were frozen to death before the morning. They were carrying the rum and they drank it. The rest of the party survived with difficulty, with the aid of a fire: a Banks greyhound, faithful beast, stayed with the poor coloured boys. Unlike Australian aborigines, they were unused to hugging dogs for blankets to keep them warm, or they might have lived.

The indefatigable Banks was no sooner back at the ship before demanding a boat to pull him around and haul the seine, but Solander, Buchan, Astronomer Green, Surgeon Monkhouse and the two Revesby footmen collapsed into their cots and slept the sleep of the dead. Cook continued with shore parties cutting wood and filling casks with water, while he surveyed the bay. With the wind against the tide he found a very nasty sea get up in which the ship sometimes threw her whole bowsprit into the water. The sea could run high at the anchorage, too, although the ship 'Road very easey with her broadside to the swell'. *How* she rolled!

Keeping close in and making all the westing possible in spite of 'fresh gales and squally weather' and a lot of poor visibility, Cook was close to the Horn a few days later: as soon as the chance of observations offered, he and Green took some lunars and worked out its position as 55° 59′ S. Latitude and Longitude 68° 13′ West. This was just one mile out in latitude and a degree out in longitude, but a degree down there is less than forty miles. It was a remarkable result, worked from data taken on a raw day of hard westerlies from the open deck of that heaving, bouncing, rolling little ship; and worked out in the reeling great cabin. Here Banks and Solander, jammed against the table, were also busily working, sorting large numbers of plants, flowers and so forth 'most of them unknown in Europe and in that consisted their whole Value', as Cook once remarked, perhaps a little waspishly. He was very tired: anchor-work and boat-work for the irrepressible botanist Banks was trying and worrying. Odd shrubbery from the ends of the earth was no cause for excitement to him. The scientists added greatly to his work and responsibilities. Hoisting out boats for them and then picking them up again, landing them and then beating the ship off and on because there was no anchorage and he had to stand in very close to nasty rocks to get them aboard again increased the risks he had to take. Risks which the sea forced on him he could deal with in his stride, but those caused by humans, though understandable, were strictly unnecessary. Just to make the passage of the Straits of le Maire and the Horn was stress enough. Banks, very obviously, never grasped the least notion of the stresses on Cook as Master. How could he? The world was his, wherever he was: the only way really to appreciate shipmaster's stress is to be Master oneself. Cook was well aware that he could wreck the ship in a flash, at almost any time.

Now he was past the Horn, and by the wind's clemency and his own skill was not blown back again, he stood steadily to the south-west and south as far as 60° South on 74° 10′ West, some 600 miles from Cape Horn: and there was no land, nor sign of land. When the gales blew from any point south the sting of the ice was in them and the open decks of the old North Sea battler were a cold hell. The helmsmen, barehanded at the wheel, shivered in their Fearnought coats and from high aloft the great royal albatross surveyed the unique scene. For no ship had ever reached that far south and west before. There was no sign of any land. The endless roll of the menacing seas raging for ever from all points west obviously came with a very long fetch. A seaman could see that there was no continental land to wind'ard there, not for thousands of miles.

It was enough. They had been five weeks off the pitch of the Horn. On 31 January 1769, the wind went to W.S.W., fell calm for a while, then backed to a moderate breeze from E.S.E. This was luck. Cook knew what to do with it. He began to haul a little to the Nor'ard and gave her all sail. 'At 3 a.m. wind at E.S.E. a Modarate breeze; set the Studding sails, and soon after 2 birds like Penguins was seen by the Mate of the watch.' Come up to see the amazing sight of a sailing-ship under stuns'ls, no doubt, the penguins barked with astonishment and sped away underwater as fast as they could go. It was as well they did. In the morning, Mr Banks was out rowing round the ship in a 'Lighterman's skiff' shooting birds for specimens – any birds.

A few days later, on 12 February, the *Endeavour* sailed across 50° South on near 90° West in the Pacific Ocean, thirty-eight days from 50° South in the Atlantic.

This was very good going indeed, and astonishing luck: on that whole passage Cook had not once been forced to get her under close reefs, once clear of the Straits of le Maire. This, he wrote, was 'a circumstance which perhaps never happen'd before to any Ship in those seas so much dreaded for hard gales of wind, insomuch that the doubling of Cape Horn is thought by some to be a mighty thing and others to this Day prefer the Straits of Magellan ...'

The doubling of the Horn was a mighty thing indeed, and remained so until the end of the sailing-ship era. The competent, watchful, stick-at-it Yorkshireman came round with studding-sails. Good Lord! The man set studding-sails off the pitch of the Horn.

CHAPTER SEVEN

A Seaman in the South Pacific

SAFELY away from the precincts of Cape Horn, Cook continued to stand towards the north-west in higher latitudes than the South Pacific had been sailed before. His instructions did not require him to search for *Terra Australis* particularly at that stage of the voyage but to reach King George's Island (Tahiti) in ample time to prepare for the observation of the transit of Venus, and look for the continent afterwards. It was still summer: Cook was not the man to waste it. Others might make a dash for the north and warm, trade winds: the *Endeavour* had not a sick man aboard, no sign of scurvy, plenty of wild celery and other anti-scorbutics, and recently shipped fresh water. Nor had she suffered the more usual wearying battle with the Straits or the Horn. Almost as if they respected Cook's temerity the sea gods had allowed her to pass lightly by. There was still plenty of hard wind – 'fresh gales and clowdy', 'heavy squalls of wind and rain', 'split the main topsail, bent another', and so on, says the log – but there was no serious damage aloft or below. The ship like others of her time was really lightly sparred, but as the result of superb seamanship and constant care not a yard was sprung. Whenever tops'ls were reefed Cook had the topgallant yards above them sent down to the deck, to ease the strain aloft, and they were not sent up again until all the reefs were shaken out.

His seamen indeed sent yards and topgallant and top-masts up and down with a greater facility than later-day sailors handled only the sails. He had plenty of men and he looked after them: they were then on the two-watch system, twenty A.B.s on deck at a time with good mates and petty officers in each watch. The *Endeavour* was far better off at that stage of her long voyage than any of her predecessors had been. This was Cook's doing – the result of his policies right through the voyage – first in selecting that tough, rough, sea-kindly hull; then in always looking after it and the people who served it; then boldly taking on Cape Horn, slipping around inshore and be damned to it. No matter how much the ship rolled and pitched – often the sea smashed over the quarter-deck where the absence of bulwarks, if uncomfortable, allowed the water at least to clear itself quickly: a couple of sharp rolls of the fat hull like a poodle shaking itself and the sea was gone – the cleaning and airing routines went regularly on. Cook saw to it that the men were always clean and as dry as possible. In really nasty weather with squalls of rain and sleet driving over her, he had canvas dodgers rigged up along the weather rail and in the lower mizzen shrouds to protect the helmsmen and the essential watch-keepers, while the other seamen necessarily on deck huddled in their Fearnought coats in the lee of the boats. These boats made a substantial wind-break. Sail was handled as far as possible before rain-squalls struck, as the weather-proof coats blew over the men's heads when they were out on the footropes with rain and hail lashing at them. They had no proper oilskin suits for work aloft, for such things had not been developed.

Regularly at least twice a week and more often if necessary, the quarters and the hold were 'cured with fires'. The ship still had a stock of north-country coal and charcoal, and fires were made with this carried about in iron pots. The fires set up movements in the air.

'I used frequently a fire made in an iron pot at the bottom of the well' – the pump-well, deep below – 'which was of great use in purifying the air in the lower parts of the ship. To this, and to cleanliness, as well in the ship as among her people, too great attention cannot be paid,' Cook wrote after the voyage. There was little or no fire risk: the fires were low, for their job was to set up circulation in the trapped air below and get the foul air out. There were plenty of reliable men to watch the fires. Wooden ships made some water through leakage, and some of the provisions of those days inevitably went bad in time. In most ships, bilge water stank to heaven and fouled the air the men must sleep in. But not with Cook. He stood no nonsense ever, not even from bilge-water.

Of course, a ship like that with a crowd of men aboard had to be organized all the time. She had a powder magazine below (Cook mixed powder with vinegar and 'cured' the woodwork with that, too) to add to the risks, but she was well guarded. Her marines stood sentinel duty by turns throughout the twenty-four hours in key places – this was one of the chief reasons for carrying them – and there were constant, and searching, 'rounds' carried out regularly and irregularly. The marines did not go aloft or perform any seaman's duties. They had enough of their own. As the ship made her way slowly towards the islands they put in considerable time at small-arms practice. Captain Wallis had had to resort to a little gunnery before the *Dolphin* was secure at Tahiti, paradise as the place was alleged to be.

Cook was taking no chances with *anything*. As the ship made northing and the sting went out of wind and sea, he had six of the carriage guns brought on deck and set up in the waist, three each side. They were lashed down to their ringbolts to prevent them taking charge and careering about the decks. The wind came nicely away from the south-east and east-south-east in March, on 33° South (which was early and premature) and the ship was put at once on the three-watch system as she had been before the Straits of le Maire. This the sailors greatly appreciated. Cook kept his ship going along under all the sail she could carry day and night, though Wallis had snugged the *Dolphin* down at sunset and taken the way off her until dawn. Wallis was a sick man on his Pacific voyage and the *Dolphin*, after all, sailed two feet to the *Endeavour*'s one. The little cat trod warily across the great Pacific sea, but she kept going – best days' runs she logged between Horn and Tahiti were 132 and 140 miles on consecutive days in February, with good south-westerly winds – and gave herself time to haul away from any dangers that might fall in her path. Already Cook was establishing that odd reputation of having a most fortunate sixth sense, the ability to know when danger was near though even the most vigilant watch-keepers had not suspected it. Up he would come on deck at such times and be ready at once to cope.

In fact, many good sailing-ship Masters developed this faculty which had its base in their complete harmony with their ships. Without engine noise or vibration to upset their senses, able to hear through the wooden (or steel) hull beside their bunks, noting the ship's roll and pitch and any change in the pattern of either, listening to the sound of the wind in the rigging, the ship's creak and the sails' play, they slept only while the pattern of all these salient things maintained its familiar harmony. An increase in the sound of the wind, a heavier roll, a pronounced swishing of the sea tumbling past the ship and they were on deck in an instant. As for awareness of the proximity of land – well, a prudent shipmaster came up on deck whenever his bunk rolled him over, which could be often if he was not sleeping soundly. And that was often, too.

A good lookout was kept at all times, particularly for signs of land. But they saw no land. The heavy seas with the south-westerly wind, which persisted for several days after the wind eased, convinced Cook that there could be no land to windward for many miles: but he continued the search. The transit of Venus was due on 3 June and he had to arrive six weeks or a month beforehand, to set up an observation post and a fort to protect it. He had plenty of time.

The heavy gear on the sails was rigged down and sent below when the trade wind came, and the best topsails and courses shifted for sails of lighter canvas. The sailmaker got on with repairing the heavy sails for their next usage (which would come soon enough, in New Zealand waters) while the seamen cleaned and painted the boats, and got up the best cables and 'wormed' them – that is, protected the hempen strands against chafe on rock and coral (as much as possible) by twisting lengths of smaller hemp along the lay of the cable. This was no great protection but it was all they could do: at least it slightly lessened the chances of a piece of saw-tooth coral sticking into the cable and cutting right through. Whatever difficulties could be foreseen Cook tried to foresee; whatever he could cause to be done about them, he did. That was the way he had been brought up in Whitby ships, and that was the way he stayed.

Whenever there was an opportunity to fix the ship's longitude by lunars, Cook and Green got out their quadrants and made all the sights they could – always a long and difficult job. If they could see sun and moon at the same time, in such a way as offered good 'cuts', or a fixed star on a clear evening (with a nicely defined horizon) looked like being passed observably by the moon, they balanced themselves on the rolling deck and manipulated their quadrants with skill and determination. A lurch of the ship might easily catch them off balance and send them flying. Midshipmen stood by to assist by hurriedly grasping the instruments and the observers as necessary to save them in a heavy roll. Cook found a favourite stance with one long arm thrust through the mizzen rigging, to lessen the chances of being dislodged.

All this was valuable and most useful, but they had no real means of knowing how accurate their results might be. Nor could they be at all certain of the longitude of Tahiti. Wallis worked lunars, too, and did his best: how good was that? Finding another navigator's discovery on the basis of the data he brought home was a new thing: the few island groups which had been found in the Pacific – the Marquesas, the Solomons, Espiritu Santo and a few more – were promptly lost again. When the *Endeavour* sailed, their alleged positions decorated a few optimistic maps, but it was not even known usually whether they were islands or parts of the coasts of the same mainland. The only way to be sure to come on Wallis's Tahiti was to get in its latitude at least several hundred miles to the east of it, and sail on that latitude until it came in sight, for the latitude was reasonably well calculated.

This Cook did. The *Endeavour* entered the tropics on 25 March 1769, on longitude 128° W., which was some 3,000 miles west of the island of Juan Fernandez. The winds were westerly, the weather 'dark and clowdy', not at all the azure trade winds of the South Seas. (East of Tahiti among the Tuamotu group, these are seasonal, too.) One of the 'people' – crew – reported seeing a log of wood drift by in the smooth sea, but the morning brought no sign of land. They were not very far from some of the Tuamotus, but that is an atoll archipelago of low coral islands, hard to see. Cook wasted no time beating about 'searching for what I was not sure to find', and stood on towards the latitude of Tahiti. There were birds about like 'Egg Birds', small white gulls which Wallis had reported and given that odd name as it had been bestowed on them 'by Sailors in the Gulph of Florida'. Such birds did not go far from land: Cook thought the *Endeavour* was 'not far from those Islands discover'd by Quiros in 1606 ... from the Birds & ca. we have seen for these 2 or 3 days past.' He knew about Quiros from a book aboard given to Banks by Dalrymple, to whom the Portuguese was one of the great Pacific prophets (as indeed he was). Quiros reported signs of a continent not far from the *Endeavour*'s track and this, with other like indications, was shown on a large pull-out chart in the Dalrymple book. Now Cook saw a few 'signs' too, which he took for what they were worth. The wind turned to the east'ard and he sailed on, noting in passing that the Madeira wine had at last run out and the great cabin had to change to grog. Despite the marine sentries, it is obvious from sundry journals that various thirsty old lags among the mariners had found ways of quietly broaching some of the gentlemen's private barrels. Several were found to be half-empty when brought up from the hold, though they were filled aboard from bulk and there had been no natural leakage.

A few days later, Master Peter Briscoe, footman to Mr Banks, high in the fore rigging to see what he could see, sighted their first island ('to ye honour of ye 2d watch then upon deck'). The trade wind was blowing fair and fresh, the weather beautifully clear, the little cat foaming along at her best speed in the perfect conditions. From high aloft Briscoe could see the whole island, a ring of palm-dotted coral reef with paler light reflected from its shallow interior lagoon, and the coconut palms waving in the wind. From deck they could soon see the palm-tops too, which looked at first as if they were growing in the sea. Cook hauled the wind and sailed as close as he could without hazard, and there was excitement among the gentlemen and the mariners. The *Dolphin* veterans had been yarning for months about the fleshly attractions of George III's Island and the assorted delights offering there for mariners, for a small nail would provide a succulent young wench for the night (or day: their co-operation was described as boundless) and a spike finance a harem for

H.M.S. Endeavour, *Captain J. Cook, in a fresh wind*

a month. This was heady stuff for lusty young men kept in such rude health by their dour commander.

But here there was no anchorage, no way into the lagoon for anything larger than a canoe, and no welcome either. The inhabitants, writes Cook, 'march'd along the shore abreast of the Ship with long clubs in their hands as tho they ment to oppose our landing'. The women hid among the palms.

Cook called the atoll 'Lagoon Island' because of the lagoon, a very familiar feature in the South Sea islands. It is on the charts now as Vahitahi: the lagoon, like others in the Tuamotus, is noted for pearls and good shell, and the coconut palms produce copra. The *Endeavour* now sailed from atoll to atoll in the dangerous group – in sailing-ship days the Tuamotus were known as the Dangerous Archipelago or the Isles of Danger, for they are all very low, rising abruptly from the sea bed so that soundings give no warning of them, and unpredictable and varying sets and currents race among them – looking for anchorage and finding none, and finding little sign of welcome either. When there were inhabitants they could be seen bearing clubs and 'long pikes'. At Ravahere the islanders prepared to launch canoes: Cook backed the main yards to take the ship's way off and wait for them. He got presents ready, but when he hove-to they stopped.

The high mountains of Tahiti itself were in sight distantly on 11 April, eighty days from Staten Island. This was no atoll: the peaks resting on the dawn clouds seemed to reach for heaven. Near the high land the trade wind, which took the atolls in its stride, became flukey and then fell calm: the *Endeavour* dribbled on. Cook was making for the harbour that Wallis used, Matavai Bay on the north coast, where the *Dolphin* found excellent shelter by a river-mouth, largely unimpeded by reefs. It took two more days to reach this. Canoes of Tahitians came out in the fine weather, but warily. None would come aboard, though they carried the green bough of peace and shouted '*Taio! Taio!*' (Comrade!) in friendly welcome. Some tossed up coconuts and a sort of fruit, accepting beads thrown down to them in return.

By night the breeze was better: by dawn of 13 April the *Endeavour* was feeling her way towards Venus Point and the Matavai anchorage, both big bowers at the catheads ready with their best cables, mariners calling the depths in time-honoured chants as they sounded in the chains for'ard on both sides, the morning sun touching the big topsails, the clewed up topgallants, the hauled up courses and the jibs with a golden light of everchanging beauty, while the blue water rippled softly at the old North Sea bow. On deck seamen stood silently at their stations to down all sail. Excited boys were in the tops ready to run out on the footropes below the yards and harbour-stow the sails, grown softened in the tropics passage. Cook – ever alert, ever the foreseeing – gave his orders quietly to the helmsmen, as he conned the ship in, Mr Gore from the *Dolphin* pointing out

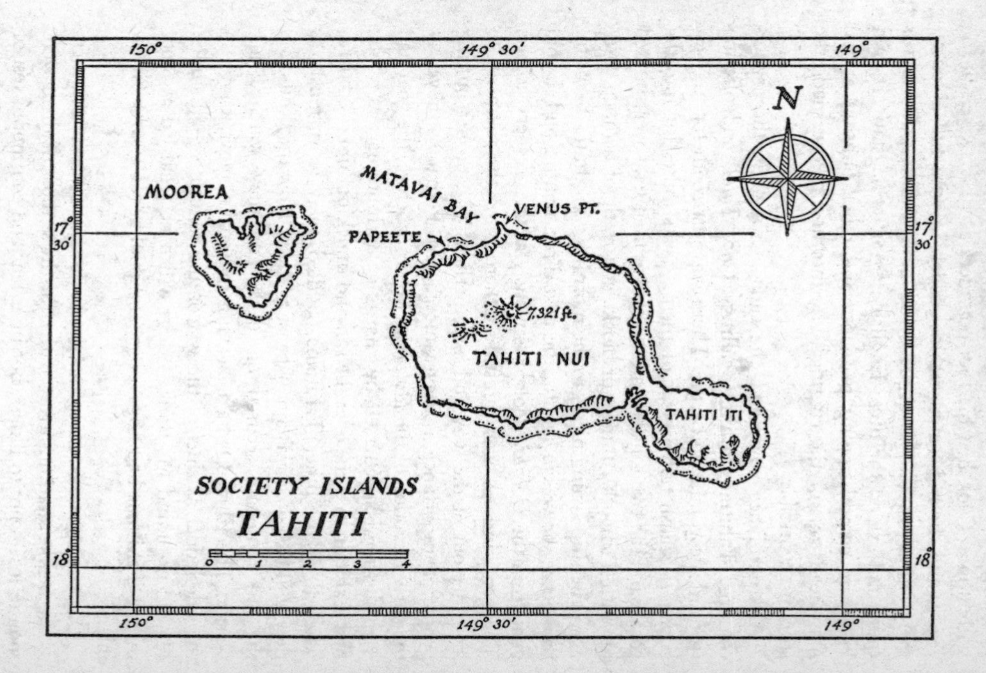
150°
149° 30'
149°
N
MOOREA
MATAVAI BAY
VENUS PT.
PAPEETE
17° 30'
7,321 ft.
TAHITI NUI
TAHITI ITI
SOCIETY ISLANDS
TAHITI
0 1 2 3 4
18°

C.C.S.S.–7

features. Another Wallis veteran was out in the pinnace to mark a dangerous submerged reef which the frigate had touched on and got off with difficulty, while the Tahitians showered her with stones. Cook had his four-pounders and his swivels at the ready, and the red-coated marines fallen in with full accoutrement and arms. There was just wind enough to keep the canvas quiet and extend the huge English flag at its flagstaff aft, and keep the commissioning pennants and other bunting at the mast-heads out of the rigging. The morning breeze made the ship just workable.

All round her now came the canoes crowding, the copper-coloured Tahitians smiling, calling '*Taio! Taio!*' and offering (to trade) coconuts and fruit. There was just room for the best bower to splash down clear of them. As its flukes bit into the coral sand the two tops'ls – all the canvas then set – were bunted up with a smooth running of blocks and rattle of hempen cordage. The ship swung nicely to the morning wind in the beautiful bay where she was to stay for the next three months, and the old goat from the *Dolphin* looked enviously at the greenery ashore. The goat, indeed, was carefully tethered, for the moment, to keep her out of the way and also, perhaps, to stop her butting people of importance as they came aboard. Perhaps annoyed at the surly reception of the *Dolphin* two years before, the goat had then distinguished herself by charging the hindquarters of the first Tahitian to come over the rail and knocking him overboard, whereat all the other Tahitians paddled away as furiously as they could. A goat was a new beast to them. Now they could see the old reprobate secured on deck, and they watched her warily. The Chief Owha'a, braving the goat's anger, came up the side ladder as chief of the welcoming committee, all smiles. He did not mind Mr Banks's dogs which he regarded as food, for the Tahitians had hogs, dogs, and poultry (of a sort), all edible.

Out swung the boats to the rhythmic shouts of the mariners at the tackles, hauling away with a lustiness most of them were soon to lose – not to illness, but in ardent and continuous test of those yarns they had heard of the extreme and glorious co-operation of the Tahitian women – or such of them, perhaps, as sailors met, who swiftly learned to value iron above virtue. The passionate pursuit and research into the arts of Polynesian love could be rough on the ship's hardware, and balance of trade. Cook took immediate steps on this. He knew that the *Dolphin*'s crew, with only one month in Tahiti, had surreptitiously extracted so many vital spikes and long bolts from her interior as to threaten her structural strength, so ardent was their enjoyment of this insatiable market. Cook had brought nails and spikes by the barrel-full, but not for the purchase of love. This hardware was strictly for the ship's needs – fish, hogs, fruit, coconuts, breadfruit, and the like – and was placed under guard. He had prepared a set of trading 'Rules' on the way up from the Horn, which were to be enforced on everyone. Banks, the Oxford man, was appointed trader and barterer-in-chief, a job in which he showed considerable talent and scrupulous fairness.

Cook's 'Rules' are worth quoting at least in part:

'I thought it very necessary that some order should be observed in trafficking with the Natives,' he wrote in his Journal, 'that such Merchandize as we had on board for that purpose might continue to bear a proper value, and not leave it to every ones own particular fancy which would not fail to bring on confusion and quarels between us and the Natives, and would infallible lessen the value of such articles as we had to traffic with ...'

So the *RULES to be Observ'd by every person*, after decreeing that all must try by every fair means to cultivate friendship with the natives and treat them with all imaginable humanity, laid down that:

1. A proper person or persons will be appointed to trade with the Natives for all manner of Provisions, Fruit and other productions of the earth: and no Officer or Seaman or other person belonging to the Ship ... shall Trade or offer to Trade for any sort of Provisions Fruit or other productions of the earth unless they have my permission so to do.

2. Every Person employ'd a Shore on any duty whatsoever is strictly to attend to the same, and if by neglect he looseth any of his Arms or working tools, or suffers them to be stole, the full Value thereof will be charged againest his pay according to the Custom of the

Navy in such cases, and he shall receive such further punishment as the nature of the offence may deserve.

3. The same penalty will be inflicted on every person who is found to embezzle, trade or offer to trade with any part of the Ships Stores of what nature soever.

4. No sort of Iron, or anything that is made of Iron, or any sort of Cloth or other usefull or necessary articles, are to be given for any other thing but provisions. – J. C.

The Rules were read out at quarters and in all the messes, and copies put up in the lobbies for'ard and aft. They all knew the 'thing' the ship's goods were not to buy. The *Dolphin* men reported that prices were rather on the high side even for the simplest articles, compared with those asked them two years earlier. A fair-sized spike would buy a hog when the *Dolphin* came, but now a hatchet was demanded, and the demand meant. Another regrettable feature they noticed was an exodus of the ruling classes who had formerly lived in the area of Matavai Bay. Only a few ruins remained of their homes. Lieutenant Gore noted that 'allmost everything was alter'd for the worse. ... A number of fine houses dispers'd among the woods, Many Inhabitants of the Better Sort, a number of Large Canoes lying round the Bay in sheds and other buildings, together with several Images of 4 and 5 men standing on each others shoulders in one piece, Hogs and Fowls about the houses ...', all were gone. He thought there must have been a revolution. The French frigate *Boudeuse* had been there between the visits of the *Dolphin* and the *Endeavour*, but Bougainville and his men had had nothing to do with all this. It must have been the 'Noble Savages' of his New Cytherea themselves.

The same noble savages whose nobility and largely imaginary social order was even then the talk of Paris, proved also to be the most ignoble thieves, 'prodigious expert' pickpockets and skilful purloiners of anything that took their fancy no matter how well guarded, as Cook quickly noticed. This became annoying. An observatory was set up ashore on ground approved by the chiefs, with a well-armed fortress and the ship's guns covering that, but the dexterous and cheerful criminals stole the stockings from beneath Mr Banks's pillow as he slept with friends in his personal tent, and anything else that took their fancy. When they also stole the valuable and irreplaceable quadrant brought from England for the special observations for which the voyage was being made, the angry Cook took forceful and immediate action. He seized some chiefs as hostages and held them until the quadrant was returned. He and Banks had a well-founded suspicion that it was the chiefs, male and female, who instigated the thefts anyway, for the 'lower orders' never kept the really valuable things they stole. These swiftly found their ways to the households of the chiefs. One thing obvious, too, about the Tahitian order of society was its excellent intelligence service. Those chiefs knew everything that went on. Seizing them generally brought swift action: the quadrant was returned but not before some fool had tinkered with its fine adjustments. Mr Spöring put things to rights again with his (or Mr Banks's) set of watchmaker's tools. But there was an obdurate chieftainess who did not return things, by name Oberea (spelled and misspelled variously by Cook, Banks and all the others who could write). To the ungallant, prosaic Cook she was 'about 40 years of age and like most of the other Women very masculine', but the artist Parkinson found her 'a fat, bouncing, goodlooking dame'. As a Polynesian chieftainess (her precise rank never was made clear) she was likely to be at least six feet high and fat and bouncy all right, for such ladies may still weigh up to 300 pounds or more. Maybe her 'masculinity' lay in her autocratic ways and failure to order delivery of stolen goods. There was almost constant trouble over thievery, sometimes with loss of life when firearms were stolen. But, on the whole, relations were good, especially with the women and girls, and the stocks of nails, spikes, and similar hardware went down and down.

On the proper day (of gloriously uninterrupted sunshine) the transit of Venus was duly observed by Cook, Green and others: several seamen received two dozen apiece with the cat-and-nine-tails for infringements of the rules of barter and harshness to the Tahitians: poor Buchan died in a fit and was greatly missed: two of the marines deserted with some of the bouncy, goodlooking dames: and Cook and party made a round trip of the island. They also learned to eat baked dog – as good as lamb, reported

Cook, resolving for the future not to despise dog's flesh, though at first he did not fancy the dish at all: local dogs were strictly vegetarian, which probably helped – and observed several blemishes in the apparently idyllic lives of the Tahitians. Much of the food, for example, was strictly seasonal, and there could be shortages. Fish though plentiful seemed hard to catch and was usually rounded up by whole villages, working in the water strenuously by night, which they did not seem to relish. The nasty disease known as Yaws was very prevalent, and there was a regrettable shortage of hogs. Few were offered for trade, and high prices (such as an axe or a hatchet, which were in short supply) asked for them. And the thievery went on and on. Only two marines – Clement Webb and Sam'l Gibson – tried to abscond. They tried hiding in the mountains with their new wives, but there was no hiding-place in Tahiti once Cook had set the chiefs to finding them. Cook simply seized four or five who were handy, including the 'masculine' Oberea: in due course the shame-faced marines came marching back, wives still in company, and the chiefs were released. It appeared that their strong attachment to two girls was the only reason for their desertion: the humane Cook, wondering that more of his crew did not prefer the island to a return to eighteenth-century England, put them in irons for the moment to reflect on their misdeeds. The wailing of their brides, so briefly wedded, rose above the palms ashore, but not for long.

The ship had been thoroughly overhauled above the waterline during the long stay – three months – in Matavai Bay, and all the provisions, stores, and everything else below landed, examined, and restowed aboard: but the sun and rain had done the masts, spars, and standing rigging no good. When they hove up the anchors their wooden stocks were found so eaten away by borer-worms that they fell off and had to be renewed. The stock of a big bower anchor was a large piece of timber: some useful replacements were found ashore, cut to size, shaped and secured in place by the carpenter and his assistants. While this work was going on and final preparations for sea continued, many Tahitians volunteered to go with the ship. Cook would not take them. It was mid-winter then in the southern latitudes where he intended to look for Dalrymple's *Terra Australis*. How much of shipboard life and food could Polynesians stand, in winter gales? From that time onwards the *Endeavour*'s would be a lonely, pioneering track – no leisured wandering in the trade wind zone. Mr Banks urged that a Tahitian pilot might at least be taken. The ancestors of these people, after all, had been making long oceanic voyages before the Vikings sailed to Vinland: they still had fleets of large sea-going canoes, and many claimed wide knowledge of other islands beyond the horizon. Among these, one seemed to be outstanding. This was a former Raiatean high priest, politician and navigator – a rather odd mixture, explained by the fact that priests tried to keep a monopoly of knowledge – named Tupia, or Tupaia, and Cook agreed eventually to accept this man. After all, he certainly knew the big islands to the west of Tahiti, for he was reared among them: he would be useful for that part of the voyage. Being so close to these islands Cook intended to sail first that way.

Mr Banks had other ideas. Mindful perhaps of the sensation that a 'noble savage' might cause in London and unaware that Bougainville already had introduced such a one in Paris, Mr Banks confided to his Journal * that he felt justified in taking the man to England 'as a curiosity', remarking that some of his neighbours kept 'lions and tygers at larger expence than he will probably ever put me to'. Poor Tupia: he did not survive to become either lion or tiger. He died of the fever at Batavia along with a small boy who came as his servant. But he was most useful while he was aboard, both as interpreter and pilot for the Society Islands.

Midmorning of 13 July 1769, holding only to a coasting anchor on a light cable, the forecastle party of the *Endeavour* shipped their handspikes to the wooden windlass. All sail was loosed: the ship lay heading into the light breeze, as ships do at anchor. The pinnace pulled ashore almost awash with the weight of Oberea and the other chiefs, lamenting quietly in

* See *The* Endeavour *Journal of Joseph Banks, 1768–1771*, edited by Dr. J. C. Beaglehole (2 Vols: Angus & Robertson, Sydney, with the Trustees of the Public Library of N.S.W., 1962).

great contrast to their subjects, who had assembled in a fleet of canoes and vied with one another in the sound and volume of their cries of grief to such an extent that the sailors had grave difficulty in hearing the orders.

But they knew their work. Getting the square-rigger under way under such conditions to them was child's play. Tops'ls were sheeted home and mast-headed, the fore aback, main and mizzen filling. As the backed fore tops'l acting as a lever swung her head around to face the open sea, the last anchor was broken out in a rush, and the little ship gathered way. Mr Banks took Tupia to the fore topmast-head to have a last look at Tahiti: the pinnace, after standing by to mark the *Dolphin* shoal, hurried after the ship, where she was swiftly hoisted aboard: the great flag of England flew above the quarter-deck brave in its sunlit red, white and blue, and the leadsmen chanting their soundings could not make themselves heard. The ever vigilant Cook watched them closely: he could see the marks on their leadlines for himself. Banks noted poor Tupia, his lion for London, brush away a tear as he stood there at the topmast head until the last canoe was out of sight. Then he quietly climbed down, and showed no more emotion.

Tupia, indeed, was very much worth taking aboard. Tahiti is perhaps the most publicized island in the whole Pacific not even excepting the professionally lauded Hawaii and Oahu, but it is only one of a group of beautiful high islands in the area. It was Cook who called these the Society Islands, for the honour of the Royal Society. Tupia's Raiatea was one of them, and he knew them all. Day after day now, Cook sailed the *Endeavour* west among them – first (leaving Tahiti's neighbour, the striking Moorea with its skyscraper silhouette) to Tetiaroa, the only atoll in this otherwise mountainous group, its lagoon inhabited by fish in vast numbers and good pearl oysters, its low land by the palm-climbing coconut crab, and large flocks of sea birds.

'A good place for fishing, and temporary refuge for chiefs from other islands in times of trouble', said Tupia, as one who knew. He advised Cook not to try anchoring here but to go on to an island distantly in sight called Huahine. This was a high island – over 2,000 feet in places – but surrounded by a reef and difficult of approach. Tupia piloted the way through a pass in the reef to anchorage off a village called Fare, where the chief lived. This gentleman very properly saw no reason to welcome Europeans or any other strangers to his island – whatever they brought could scarcely be good: he had heard of excessive demands for food in Tahiti, and a man shot dead for stealing a musket – and was inclined to take to the hills, until he saw Tupia. Then all was friendliness. Tupia offered prayer and appropriate gifts: hogs, coconuts, and fruit came back in exchange. On leaving, within a day or two, Cook gave the chief a plate on which the armourer had cut, as deeply as he could, the following:

His Britannick Maj. Ship Endeavour
Lieut. Cook Commander 16 July 1769.
Huaheine.

This was not only a gift but intended as warning to future interlopers that H.B.M. had been there first. Some Georgian coins of 1761 were also given, and other presents, and the chief swore never to part with any of them. Tupia was interpreter in all this. A request on the chief's part that the *Endeavour* should embark a party of his warriors to sail to Borabora and exterminate the inhabitants there, as the traditional enemies of Huahine, was not granted, or pressed. Instead, Cook sailed on to visit the wild men of Borabora (which he logged as Bolabola), with a call at Raiatea and Taha on the way. These islands are very close together – so close that the wind is flukey among them, and Tupia's local knowledge was of great value both with the wind (for which he prayed to the great god *Kane*) and the landmarks. Banks noted that he prayed only when he saw a catspaw coming, being a priest of long experience: the benevolent Cook, perhaps a better judge, noted no such duplicity.

Despite Tupia's now proven knowledge and capabilities, Cook always hoisted out a boat and sent the Master – Robert Molyneux, a *Dolphin* veteran – ahead to investigate harbours, taking no chances. Lookouts manned the mastheads, for coral patches showed better from up there. Leadsmen manned the chains.

Noting what he regarded as the inefficient manner of English sounding, Tupia improved upon it with a very effective method of his own. The *Endeavour*'s seamen, standing overside on the channels – the stout platforms to which the deadeyes for the lower rigging were attached to give this rigging a greater spread and therefore more strength – had to coil a marked line armed with a heavy lead, fling the lead laboriously ahead of the ship after giving it a good swing, wait until it touched bottom (if it did) and the ship sailed up to it so that the line was vertical, note the mark then nearest to water-level, and shout it with a prescribed chant. That done, they hauled the lead to the channel as quickly as they could, getting thoroughly wet (despite some protection from canvas aprons) and began the process all over again. This took time. It gave the depth only where the lead fell: pillars of coral could be anywhere. Instead of all this labour, at Huahine Tupia shouted a few words to a compatriot close by in a canoe, who immediately leapt into the sea, dived like a penguin straight beneath the keel, swam underwater to the heel of the rudder, bounced up on the quarter and shouted not only the depth but a description of the bottom, and continued to do this – Tupia interpreting – until the ship came to anchorage. This was in a very narrow channel and Cook was delighted with his human sounding machine. The 'machine', tossing his head with a great grin at the astonished faces above him each time he surfaced, seemed delighted too. After all, Tupia and all other Polynesian navigators had used these channels only for very shallow-draught canoes and now there was need of a considerable depth of water.

Noting everything and surveying all he could, fixing the positions of the islands and main headlands as he went – if only by distances and bearings from Matavai Bay, the latitude and longitude whereof had been established as well as that of any place away from the Greenwich or Paris meridians – Cook continued on his thorough way. He hoisted the 'English Jack' here and there, taking possession for George III in the prescribed manner, as first discoverer. These volcanic islands though spectacularly beautiful and promisingly productive were no *Terra Australis* nor parts thereof. Refreshments of plaintains, yams, coconuts, 'a few Hogs and fouls', were taken aboard as they offered, and sufficient wood and water. A bothersome leak in the forepeak was caulked at anchorage in Raiatea lagoon. The *Endeavour* was in good order, the mariners recovered from the exercise of their virility at Tahiti, the helpful Tupia well settled in, and Mr Banks and party accustomed again to the confinement of life aboard after three months in tents. Even the goat from the *Dolphin*, which had enjoyed the life ashore (and very probably, also the power of terrorizing the natives with a glare from those baleful eyes and a toss of her twisted horns) was comfortable again on the hard deck in the lee of the longboat.

It was time to leave the islands, haul up to the trade wind, and stand to the south. If no continent offered earlier, there was certainly the eastern coast of Tasman's Staats Land (or New Zealand), to be discovered there somewhere. It must have an east coast somewhere. Cook meant to find it.

A sting came in the air again beyond the trades, and the cold sea spray lashed the little ship once more as on the long slog up from the Horn. The mariners, used for so long to tropic warmth, put on the Fearnought coats, wrapped one round the goat, and shivered. The old goat was of tougher stuff than the few surviving hogs and poultry from the Islands, for they gave up and died. Tupia, too, suffered from the cold and complained of a pain in the stomach. The Irish Bos'n's Mate, John Reading, may have had a pain in the stomach, too. Whatever it was he treated with three half-pints of rum provided by sympathetic messmates who had saved it up. All this he drank undiluted and consecutively, as he liked the stuff. Normally this might have put him to sleep for a watch or two. Though considerable, the quantity was not beyond his capacity. But this time it was the wrong medicine and he slept for ever. Aged twenty-five.

CHAPTER EIGHT

On the Barrier Reef

'AT ½ past One a Small Boy who was at the masthead call'd out Land!' wrote Banks in his Journal for 6 October 1769. The lad was Nicholas Young, a juvenile aged twelve or so, who must have been surreptitiously aboard from Plymouth or perhaps Deptford but whose official existence began on Mr Buchan's death at Tahiti. His name then appears for the first time on the muster roll. Perhaps he was a real live 'widows' man'. There were two such wraiths in the *Endeavour*'s complement but they were meant to be names, not live bodies. Every H.M. ship at the time had to carry fictitious seamen on her roll at the rate of one for each fifty real seamen, the idea being that their pay and cost of victualling went into a naval widows' pensions fund. The purser – in the *Endeavour* this was Cook – kept their accounts and handed the money over at the end of the voyage. 'Nick' Young could have been any juvenile brought aboard with the connivance of the seamen. A handy lad was usually welcome. There were plenty of them aboard the ships at the Battle of Trafalgar thirty-five years later, and women too.

Whatever doubt there might be about the 'Small Boy', there was no doubt about the land he sighted. It was real land and no cloudbank: it was in the latitude of Tasman's New Zealand, and there seemed to be a lot of it. More and more filled the horizon in both directions as the *Endeavour* came slowly on. It took another couple of days to close the land, which Banks was already recording as 'certainly the Continent we are in search of'. The landfall was on the coast of Poverty Bay (so named by Cook) where the town of Gisborne was to grow.

The passage from Tahiti had taken over six weeks, most of it grim, the ship's behaviour 'very troublesome' to the gentlemen and trying for everyone else. She had to pound into the combers of the Roaring Forties most of the way. Crossing the zone of variable winds she had to put up with whatever came, and this was almost as trying, for the square-rigged ship was a hard task-master and took unkindly to foul winds. Cook sailed from the Society Islands straight down past 40° S., found the westerlies threatening damage to the rigging which had then been taking punishment for over a year, slipped to the nor'ard for a day or two to give men and rigging rest, then made westing in quest of the east coast of Tasman's discovery. All this area came within Dalrymple's conjectured continent. From his experience coming up from the Horn Cook was sceptical about this, and the ship's officers were with him. The subject must have been thrashed out frequently aboard: the *Endeavour*'s small library included the relevant books. Banks indicates that the great cabin club – the afterguard and the 'gentlemen' – took friendly sides, the pro-continents and the no-continents. He himself headed the pros: Cook noted their failure to find the place. There was obviously a vast area of the Pacific still to be searched, but the *Endeavour* had already sailed through a lot of Dalrymple's 'land'. Here at last was some real hard earth risen from the sea, and Cook was determined to examine it thoroughly if it took a year. Tasman, after all, had seen part only of its west coast: might it be a great peninsula jutting up from Antarctica? How far south did it extend, or south-eastwards?

First contact was not encouraging. Polynesians speaking Tupia's language came out in canoes, well armed with lances, spears, braining clubs made of stone and wooden clubs with spikes. Cook wanted nothing more than to be friends with them (and to survey their country, claiming it in due course for George III for possible future take-over: this was his duty). But these Polynesians had taken over by bloody conquest of their predecessors not so long before. Their policy was the simple one that all strangers were enemies, to be put to death if possible and eaten if they tasted good, just in case any desirable fighting qualities or knowledge could be absorbed from their flesh and broiled hearts. Having seized the country by sea themselves they had no intention of allowing others to repeat the performance, even if they came mysteriously by 'floating island'. (There was no tradition left of Tasman's ships anywhere: it was well over a century since the Hollanders had been on the other coast.) These pale strangers had 'sticks' which could strike lightning and belch fire: and their 'island' had larger apparatus of the same kind. The Maoris did not lack courage. When fire was loosed at them they fought back with defiance. In the first encounter one of these 'very obstinate and stubborn people, and brave withall' seized a sword from Astronomer Green after Cook had caused some small shot to be sent over – not at – them. Instead of flight or surrender at the guns' roar, they shouted defiance and paddled in to the attack 'in a very bold manner', greatly to the admiration of the English fighting seamen.

But they *had* to learn. Ball followed the shot: several Maoris died. Even then the others did not give up, for they fought as long as they had things to throw, even a parcel of fish they had in the canoe. ... Courage at canoe level was greatly admirable, but the outcome of such encounters was inevitable. There were no European casualties.

All this was extremely disagreeable – 'Black be the mark for ... the most disagreeable day my life has yet seen,' wrote Banks, who was the first to shout when the fracas started that the sword-snatcher must be punished, and fired at him. The Maoris did not, apparently, reproach the Englishmen, nor feel too badly about their own losses. A man's life was combat: they had plenty of warfare of their own. Some tribes were friendly. Others learned to co-operate. Many of these Maoris were indeed the 'noble savages' of the South Seas, though not in quite the manner most Europeans chose to imagine.

Wary of the natives yet trying always to be at peace with them, learning from them (through the excellent Tupia) and trading generously with them, Cook began his examination of New Zealand. He sailed first towards the south and reached within fifty miles or so of the south-eastern tip of the North Island – which of course he did not then know – before deciding at Cape Turnagain to 'bout ship and sail northwards instead. Increasing cold and adverse winds had something to do with his decision, but it was obvious good seamanlike sense to go north first. The southern summer was coming on: if Tasman's land extended much towards the Antarctic it was common sense to give it time to warm up a little. Cook knew the approximate latitude of the northern extremity: this was on Dalrymple's chart.

So he began that extraordinary circumnavigation of both islands of New Zealand which, even in his spectacular seafaring career, was an outstanding achievement. He had to sail over 2,500 miles in difficult water off a dangerous and unknown coast, much of it exposed to the Roaring Forties, extending to within a few degrees of the latitude of Cape Horn. He had to make as good a survey as he could. He had to do all this with a single ship and no means of communication. He was at the Antipodes, as far from base as a European ship could get: if he were lost no one would know where to look for him. His ship was a small square-rigged ship, lightly sparred, which had already been sailing for over a year; yet there he was, pitting her and her people against the Roaring Forties and the Tasman Sea – two notorious breeders of bad weather – and determined to stay upon that coast until he had put it on the world map no matter how long it took or what the difficulties.

A sailing-ship Master compelled to coast with the wind blowing onshore, had to keep off or risk wreck; yet if Cook did not sail close in, no one would learn much. Submerged rocks licking their ship-tearing edges just below the surface were difficult to see: ledges of them could be anywhere. To sail into what looked like harbours, even after the Master had done some sounding from a boat, meant some element of risk, and doing things which the prudent seaman most dislikes having to do. Cook was above all else a prudent seaman. He had to be audacious too: if he did not investigate harbours he would never find a base or haven to gather wild celery and other essential anti-scorbutics, or for rest for the people (Maoris permitting) and the chance to cut wood and collect water. These he had to have: the demand was almost insatiable. If the wind was offshore he could be blown away, and often was: then he must fight his way back again. If there was no wind at all the ship had to stay where she was, at anchor if possible, or drift at the mercy of current or tide. She could anchor only where the depth of sea permitted her anchors to reach the bottom, and the bottom must offer good holding.

Her cables were at best not much over a hundred fathoms long. If two were joined together end-for-end that was no real help, as they were most difficult to heave by hand aboard again, and there was no other means of recovering them – no means of taking the strain off the anchor until it broke clear.

The plain everyday difficulties of handling these ships, though the last engineless Cape Horners went out of commission only in 1950, are already so forgotten as to seem incredible. Their means of movement was the wind properly directed to their sails: yet they did not accept the mercy or the vagaries of the wind. They had to *fight* for their way, fight for their lives at times. To them lee shores were anathema and onshore winds on rocky coasts the threat of death. Cook had good boats, of course: but even with twenty stout men in each of the two best they could tow only in still water or with a favouring current. A small ship could make some progress slowly in flat water – up rivers, or in winding harbours – by what sailors called 'kedging', which meant carrying a light anchor ahead by boat and throwing it overboard, then hauling the ship up to it, then carrying another anchor further again, hauling to that, and so on. This was of no use in the open sea.

No, a sailing-ship Master compelled to work his ship even on a familiar coast must often have his heart in his mouth. When Cook came there the New Zealand coast was familiar to no man. Tasman had not even landed except once briefly on an offshore island. The *Endeavour* touched perilously on a submerged rock, brushed a cliff with her lee yardarms (fortunately the cliff was steep-to) while beating away from sudden peril caused by a shift of flukey wind, blew out some of her best sails, narrowly avoided wreck by night on a reef of offshore rocks near Stewart Island (Cook named them The Traps, with good reason) and again by day in that windy place Cook Straits, where only a desperate anchoring saved her when a strong tide pushed her broadside towards more rocks in a calm. Cook Straits had a bad name throughout the sailing-ship era.* With an odd

*In the ship *Joseph Conrad* in 1936, I made Cook Straits intending to call at Wellington, but off that windy harbour a sudden wind howled at me as I tried to beat in and put me in considerable danger of going ashore. I

touch of humour, Cook named one stormy headland on the northern coast, Cape Brett in honour of Admiral Brett, because it had a large rock pierced right through and the Admiral's name was Piercey. The pierced headland, affronted at this, whistled up nasty head-winds for the following week, and Cape Maria van Diemen summoned up more. The *Endeavour* took several weeks to make about a hundred miles. (The wretched places were still at it when I passed that way in various square-rigged ships in the 1920s: old Sir Piercey's pierced headland was a foul place to weather. Just past it to the north was the lovely Bay of Islands, and that was glorious.)

It took over six months to chart the coasts of both islands. Cook found attractive harbours like Ship Cove in Queen Charlotte's Sound, careened the ship (that is, grounded her in a safe place with an adequate rise and fall of tide, emptied her, hauled her over first one way and then the other by masthead tackles rigged to big trees and convenient rocks) investigated and 'payed' (covered with an anti-fouling mixture) the whole bottom. He made friends with many pleasant Maoris, for they were a gentlemanly race though warlike. These friends remembered him while they lived and were handing down traditions of him a hundred years afterwards – like that recorded by a chief in the Mercury Bay area, by name Hore-ta-te-taniwha, who met Cook and his men when a small boy. At first all the Maoris were sure the Britishers were some sort of supernatural beings because they rowed ashore facing aft and so could see where they were going through the backs of their heads. Besides, they could remove their skin (their coats) and some their scalps as well (their wigs). Later, the small boys visited aboard.

'There was one supreme man in the ship,' the chief remembered fifty years afterwards. 'We knew that he was chief of the whole by his perfect gentlemanly and noble demeanour. He seldom spoke, but some of his 'goblins' spoke much. He . . . came to us and patted our cheeks and gently touched our heads. . . .

---

wore away, and, taking no chances, stood out to sea again with a rising gale, and sailed on non-stop to Tahiti. We had had trouble in 1929 getting through the strait with the ship *Grace Harwar* bound from Australia towards the Horn.

My companions said: "This is the leader, which is proved by his kindness to us; and he is also very fond of children. A noble man cannot be lost in a crowd".' *

Except in so-called civilization, of course. That man so sure of his own 'nobility' (for which he was not to be blamed in the eighteenth century), Mr Banks, grew a little testive, perhaps, during the long circumnavigation of New Zealand. It is obvious that he and his party, on the whole, had fitted in very well: there is little evidence of friction in any of the journals. These were Service documents and not meant for personal reflections: their absence of human interest is regrettable now, for so little is really known about the personality of that reticent *rangatira* James Cook. But sometimes between the lines of a few pages – very few – there is unmistakable sign of the strain the naturalists could be to him. Though a good shipmate, Banks obviously regarded the farm labourer's son as his personal ship-driver, at least subconsciously – a skilled ship-driver, and interesting: but to Banks the ship and the voyage were fundamentally for J. Banks, Esq., of Revesby Abbey, Soho Square, Oxford, the Royal Society, etc. It was the attitude of eighteenth-century landed gentry.

At any rate, on 14 March 1770 the *Endeavour* was bowling along, for once with a fair wind, off the south-western coast of the south island of New Zealand. It was a bright clear day, and the westerly wind sang busily in the rigging as the little ship rolled and pitched easily on her way. She was off a particularly lovely part of that beautiful island, and the sun touched the snow upon the high mountains to a shining glory. They were close enough in for the scientists to see lights glinting and dancing in some odd veins of the exposed coastal cliffs. What ores might be found there? It was some days since they had stretched their legs ashore. There was a crack between the cliffs leading into a valley between the mountains, a sort of deep fjord. It looked a fascinating place to land. Banks and Solander wanted Cook to go in, regardless. But he could not see from outside whether

* This incident was recorded from the chief by John White and appears in his *Ancient History of the Maoris*, Vol. 5. The Maoris could appreciate a true aristocrat.

there really was a useful harbour: whatever else it offered it was most unlikely to be a good anchorage. Fjord waters are deep: fjord bottoms are often rocky, poor stuff for the flukes of clumsy anchors: fjord cliffs funnel hard winds. The fresh wind, rushing at the cliffs, blew straight in. If he turned the ship shorewards she would rush headlong with them, even with minimum sail: what then, if there was no good anchorage? To a seaman, it was a fool place to sail into, a stupid and unnecessary risk to take. The whole of this west coast of the south island was a lee shore, dangerous to sailing-ships. Hard winds blew them on, not off. How could Cook know that the coast, still out of sight, might not reach out towards the west almost at any moment and embay him? Nobody knew the answer to that: he had to find it out for himself. The acceptance of avoidable risk was not audacity but irresponsibility.

So Cook refused. He would do all he could to meet his passengers' wishes, but he would not hazard the ship.

There were words – just what words we do not know. But both men wrote of the incident in their journals, and Banks remembered it. He was not used to being thwarted.

Here are Cook's observations from his Journal:

'The Land on each side of the harbour riseth almost perpendicular from the Sea to a very considerable height and this was the reason why I did not attempt to go in with the Ship because I saw clearly that no winds could blow there but what was either right in or right out. That is Westerly or Easterly, and it certainly would have been highly imprudent in me to have put into a place where we could not have got out but with a wind that we have found lately does not blow one day in a month: I mention this because there were some on board who wanted me to harbour at any rate without in the least considering either the present or future concequences.'

It is obvious who the 'some on board' were. How far did Banks and Solander indicate their displeasure? Banks logs his regret at missing the landing: he never forgot the incident. More than thirty years afterwards, when Cook had been dead for quarter of a century, he was still annoyed about it. Praising some diversion made by Matthew Flinders for the benefit of scientists, he compares Cook unfavourably with his fellow county-man, the little Lincolnshire navigator who deserved, he wrote, 'great credit for the pains he must have taken to give a variety of opportunities of landing and botanizing. Had Cook paid the same attention to the Naturalists, we should have done more ...'

This was stupid comment and somewhat petty. Cook forgot about the slight affair as soon as he had written up his journal. He was carrying the *whole* responsibility for the *Endeavour* and her voyage: neither Banks nor Solander had much, if any, idea of just what this meant. Cook was a short-tempered master-mariner: how severely, one may wonder, did he tell his scientists off on the quarterdeck that day? Enough, obviously, to feel that he must explain himself to his journal: certainly unforgettably. Banks was not a vindictive man though gouty by the turn of the nineteenth century. (He had then been the dictatorial president of the Royal Society since 1778.)

Within a week or two they had completed the survey and examination of New Zealand; Cook made again for the pleasant area of Queen Charlotte's Sound on the north-eastern end of the south island, just across Cook Straits from Wellington. Now he was entitled under his instructions from Admiralty to go home. What way was best? The ship was still in as good order as a small wooden square-rigger could be after so long and hard a service. No one had scurvy. There were still large supplies of anti-scorbutics aboard, and she was full to capacity with firewood and good fresh water. She had stores enough. So, 'being now resolved to quit this country and ... return home by such a rout as might conduce most to the advantage of the service I am upon', Cook called a council of the afterguard to decide the 'most eligible way of putting this in execution'.

He wished, personally, to run for the Horn and reach the Atlantic that way, 'because by this rout we should be able to prove the existence or non-existence of a Southern Continent which yet remains doubtfull'. There was indeed a great unknown area of the South Pacific between 40° and 60° South which he had been unable to sail into. But the season was late for Cape Horn. It would soon be winter. Many of the sails had been damaged and the spars and rigging had taken heavy punishment. The council of seamen thought it unwise to try the Horn again that voyage. Cook accepted this sensible view. It was decided instead to sail westwards over the Tasman Sea until they came upon the east coast of New Holland. It must be there somewhere. Then they would turn north following the coast, and make for the East Indies by whatever way might offer.

Tasman had crossed the sea named after him from the south of Tasmania to the north-western end of the south island of New Zealand. Now Cook, taking his departure from the same place, reversed that passage. It was tough either way. The Tasman sea stands wide open to the Antarctic, and bad-tempered winds funnel up howling with the cold or roar across from the west'ard screaming in protest at their own excessive speed as they go. In later sailing-ship days we called the Antarctic winds 'Southerly Busters' because they 'bust' upon us so rudely: the westerlies were too frequent to have a name at all. They blew right in the *Endeavour*'s face, and the southerlies pushed her north. Gales 'fresh', 'brisk' and strong strained the sails, and split the mizzen tops'l, while forcing the ship to the north, off course. The sprits'l-top'sl was so 'wore to rags' that Cook condemned it and took its few good cloths to repair his two topgallantsails, which were getting threadbare. The sprits'l-tops'l was a sort of hangover from the days when a small mast was stepped cumbrously on the end of the bowsprit and a sail set from a light yard which hoisted on it. This surely was the most optimistic, and useless, piece of canvas ever spread from a square-rigged ship, not excepting Jemmy-Greens, water-sails, 'moonrakers' and the like sometimes used but more often talked about in the clippers. Its successor in Cook's time, this sprits'l-tops'l, was little if any better, except that being set from a light yard (or 'sprit') hauled out along the jib-boom and sheeted to the arms of the spritsail-yard inboard of it on the bowsprit, it did not strain the headgear so much.

The *Endeavour* had a couple of jibs as well as the sprits'l and sprits'l-tops'l. Being fore-and-aft sails either set on hempen stays between the fore top-mast head and bowsprit or jib-boom, or hoisted and set on the strength of their own bolt-ropes, jibs were not wholly satisfactory until the days of iron wire rigging. With fresh side winds or head winds, they could put heavy strains on the head-gear. The old-time mariners were a conservative lot: down the centuries they had learned to rely on the spritsail as one of the best manoeuvring sails aboard, especially in the days of high, built-up aftercastles, for then it offset the appalling windage of such constructions and balanced the ship.* It was also a grand sail to run the ship off, being big, manoeuvrable, and so far for'ard. So Cook kept the sprits'l and the jibs, and discarded the useless sail beyond them. All his sails were made of flax canvas – good stuff, strongly woven, but not able to accept the wind's vagaries and the calm's chafe for ever. The topgallants'ls and the sprits'l-tops'l were sewn of very light canvas.

All this time a good lookout was kept: Tasman's Van Diemen's Land could be an island – nobody knew – and it was possible, perhaps, to underpass New Holland and come in on some bulge of the south coast instead of the east. A land bird 'was seen to pearch in the rigging' while the ship was under close-reefed tops'ls in a W.S.W. gale on the sixteenth day at sea: Cook sounded with his best deepsea lead but got no bottom at 120 fathoms. The following day various short-range birds were seen, among them the Tasmanian mutton-bird possibly from the Bass Strait islands (where they flourish). Cook's lunars already put him a degree to the *west* of the Tasmanian east coast as charted by Tasman: the *Endeavour* was in fact making towards Bass Straits. Cook 'brought to' – shortened sail and backed the mainyards to take the way off, in order that the ship might make minimum way† – sounded (no bottom at 130 fathoms) and waited for the morning.

And there, on 20 April 1770, was the south-eastern corner of Australia. To the south'ard was no sign of Van Diemen's Land

*I learned all this when sailing the replica *Mayflower* from the Channel to Plymouth, Mass., in 1957. She would not have manoeuvred at all without the spritsail.

† 'Bring-to, to check the course of a ship when she is advancing by arranging the sails in such a manner that they will counter-act each other, and prevent her either from retreating or moving forward': Falconer, 1780.

(for the sufficient reason that it was not there but well below the horizon): Cook got the t'gallant yards aloft, set all sail, and made off to examine this new coast wherever it might take him. Lieutenant Hicks, in charge of the watch, was the first to sight the land but there was no gallon of rum for him as there had been for the first to sight New Zealand, nor even a pint. Perhaps Cook looked on this as not really a discovery, for Tasman had led him there.

Tasman led him to the east coast of New Zealand, too: but there was a chance that the land first sighted there *could* have been an extension of *Terra Australis*. There was a curious unanimity that the only real continent in the South Pacific and South Indian oceans, whatever else it was, could not be Dalrymple's counter-poising *Terra Australis*, perhaps because it could not be large enough after Tasman's first voyage had pinned it down, or because its west, north-west, and south-west coasts were all considered so barren and useless. *Terra Australis*, to be of use, had to be big, rich, and readily exploitable.

The headland in the land-mass first sighted was named Point Hicks but, for once, there was some ambiguity about this precise point. Cook described it as 'the Southernmost Point of Land we had in sight, being at this time in the Latitude of 37° 58′ S. and Long'd of 211° 39′ W.' (148° 21′ E.). This was the ship's position. He judged the point to be on 38° S., 148° 53′ E. Today Cape Everard is in about this position and Ram Head – another of Cook's names – is some miles to the east. Both places are on the extreme south-eastern coast of the state of Victoria, near Cape Howe, which is the south-eastern extremity of Australia. From this point, Cook stood to the nor'ard along the east coast of this 'last sea-thing dredged by Sailor Time' which stretched away before him, waiting for a discoverer, as far as Cape York. Any seaman with courage enough to leave the tropics of the Pacific and sail west in the temperate zone *could* have found it. It was odd that it should have had to wait so long.

Away to the north the little *Endeavour* sailed, Cook and the other navigators busy with a running survey, noting much, naming much, and – understandably – missing much, too, especially useful harbours liked Twofold and Jervis Bays, and

Sydney Harbour. The coast Cook sailed along was almost 2,000 miles in length and he had then been over twenty months at sea. All strange coasts were dangerous: the only way really to examine them was by using boats and going frequently ashore to set up instruments. On a coast like that of New Holland (this part now includes the states of New South Wales and Queensland) to do this would take years. A 'large hollow sea from ye SE rowling in upon the land which beat everywere very high upon the Shore' was no help: the ship had had these conditions 'ever sence we came upon the Coast', Cook laments, passing the harbour which later became the naval base of Jervis Bay. Wind and sea discouraged him from entering. So he chose Botany Bay (first called Stingray Bay, then Botanists' Bay) which is not a good harbour at all. Its entrance was obvious; no sea broke across, and the Master reported a clear way in. The wind was blowing right out of the bay, which was a good sign for getting away again: the *Endeavour* sailed in without trouble and Cook anchored her 'under the South shore about 2 mile within the entrance in 6 fathoms water', tolerable holding ground. Taking Mr Banks, Dr Solander, and the seafaring Polynesian priest Tupia with him, Cook landed, and made for a place where several Aborigines had assembled. These all made off, 'except two who seem'd resolved to oppose our landing'. Not wanting bloodshed among these poor Stone Age fellows, Cook lay off: but Tupia could understand nothing these primitives said. All that was obvious was their poverty and determination to defend the few small bark huts which were their homes. Useful contact seemed quite hopeless, even to Banks. Beads, cloth, nails and all such trinkets meant nothing to them. Neither did the strange white men and their extraordinary ship. Could this be some apparition from the dim past of their 'Dreaming Time', a ghost out of aboriginal mythology? (It was to be a century or more before Europeans grasped that there *was* any aboriginal mythology. These 'blacks' looked like Dampier's men, except that their land was better and they were not so greatly bothered with flies.)

Cook raised the flag, took possession of the continent for King George, caught fish, replenished fresh water and, with Banks, formed a rather optimistic opinion of this Botany Bay. The aborigines kept out of the way, except when they had to use their very primitive canoes – 'the worst I think I ever saw,' said Cook – to fish. Banks and Solander were enraptured with the many strange plants, the beautiful birds, and the giant gum-trees: but all attempts to learn anything from the aborigines failed. A suitable inscription was cut into a large gum tree; the English colours were 'displayed ashore' every day with the idea of making the natives familiar with them; a sailor named Forby Sutherland, thirty-one years old, hailing from the Orkney Islands, died of a consumption caught from sleeping wet off the Horn and was buried at the watering-place; and Cook, having failed at all attempts to establish friendly contact with the natives, hove up anchor and sailed. A few miles to the north he noted the entrance to Sydney Harbour 'wherein there appear'd to be safe anchorage'. Indeed there was enough room for the whole British fleet: Cook was two or three miles off, the ship enjoying for once 'serene pleasant weather'. So he sailed on. He could not investigate everywhere, obviously, and Banks and Solander already had the great cabin and much of the hold full of specimens from their beloved Botany Bay.

On sailed the little ship well offshore but close enough to make a good coastal survey. Winds were often southerly by day and offshore by night, gentle breezes and clear weather – ideal conditions. She approached the tropics under studding-sails. The weather was not uniformly good but, compared with the long circumnavigation of New Zealand, coasting along by New South Wales was almost easy.

There was a price to pay. As he passed further north, off Queensland, Cook got into the labyrinth of the Great Barrier Reef – a '*coste dangerieux*' indeed, as the mysterious old chart described it. He sailed north inside the reefs with the utmost caution, sounding ahead with boats, sounding aboard from the channels, while masthead lookouts watched the water. The great reef at first gave the coast a reasonable berth and grew offshore; but it was a fantastic navigational hazard for the Whitby cat and the Whitby captain to sail into and try to survive. The Great Barrier Reef is not just a reef or a line of reefs but a horrid huge zone of them – 80,000 square miles of the age-old remains of the too industrious and infinitely too abundant coral polyp, stretching from below the Tropic of Capricorn to the coast of New Guinea in a 1,200-mile line, sometimes a hundred miles wide. Further north, past Cairns on the North Queensland coast, the ghastly polyps have left their minute calcium bones by the billion billions close inshore, in serried lines or by the square mile, waiting below the sunlit surface to bite at ships. Their bite is hard and often fatal.

After making the best part of a thousand miles in this sort of thing, sounding, examining, checking, surveying all the time, suddenly when she was approaching the end of the reefs area, the *Endeavour* struck. It was in the middle of a peacefully beautiful moonlit night, with the moon throwing brilliant light from the close-reefed tops'ls – close-reefed to give minimum speed and still allow the ship to be manageable – and touching the slow ripples from the broad bow to a silvery radiance: a beautiful tropic night – a lulling night, perhaps, for watch-keepers grown long used to intense vigilance. Sounding had just given 17 fathoms – over a hundred feet. There was no sign of danger. Before the leadsman could haul the lead up for another sounding she touched with the horrid, rasping sound of keel on coral, the sudden alarming shock of the floating ship upon the instant stopped and dead. Cook had just turned into the cot in his sleeping cabin. He was up in a flash 'in his drawers', taking cool and competent command. This was a situation he had not known; but he had foreseen it, worked out what could be done. All hands came tumbling up, the watch below rubbing their eyes a moment, though they knew what had happened. They had felt the shock and heard the noise in their hammocks.

Clew up everything! Aloft and stow!

Get the sails off the ship in order that they could cause no further damage, driving the keel along the coral, ripping out the bottom.

Bare-foot mariners raced aloft, ships' boys and midshipmen with them. Master and Mates sounded round the stranded hull to test in which direction (if any) she might best come off. As they sounded it became obvious that the tide was dropping: they had struck on the top of high water, or close to it.

The ship lay silent, pranged upon the coral, but fortunately working very little, as yet making no water. But soundings under the larboard bow where she had struck showed four *feet*: elsewhere there were three and four fathoms.

Hoist out the boats! Mariners more familiar with their ship and her gear than with their own homes ran to rig tackles, brace fore and main yards, hoist out the pinnace, the yawl, the big longboat, using their own manpower aided by the capstan and windlass, all with speed and efficiency. They needed the boats to run out anchors on long cables, to heave the ship off when the tide made again. Few orders were necessary. All knew their work. Out went the boats, out the stream anchor, overboard a couple of lighter spars to sling the stream anchor beneath the longboat and run it out astern: back came the longboat, paying out cable: off with another anchor and yet another – both bowers, both coasting anchors, all the cables. Carry them out and place them swiftly but carefully.

As the boats dropped these anchors at the extremities of the cables they came back to the ship and the mariners tumbled aboard. Now they swarmed aloft again, on all three masts together, this time to send down the topgallants and tops'l yards, the course yards, the topgallant masts and topmasts, stripping the ship right down to her lower-masts like a hulk – all this in the brilliant moonlight while the water fell and the ship began to roll and work a little – very little, for that flat bottom Cook had chosen showed its worth now. (The *Dolphin* would have fallen over and become an utter wreck, as H.M.S. *Pandora* did further north when coming back with the 'mutineers' captured from Bligh's *Bounty* at Tahiti.)

Daylight came racing in as if switched on, and the weather was still glorious. Midmorning, at high water – to Cook's consternation, a foot or more below the level of the night's – they hove on all the anchors, with the cables taken to the capstan (she had only one) and windlass, but all the heaving made no difference. The *Endeavour* would not budge. They had already tossed overboard all they could to lighten her – even six of the guns and their carriages (buoyed for later recovery), and 'iron and stone ballast, casks, hoops, staves, oyle jars, decay'd stores', these last being carried dutifully home to be surveyed and taken off ship's charge. The rigged-down masts and spars were lashed in the water alongside, the sails made up and stowed below, the boats lying ready.

The tide began to ebb again: she did not show the slightest sign of floating or coming off. They rigged tackles to the cables and bowsed them tightly in to keep the ship where she was. To get off she had to *float*. Other movement, if possible, was to be avoided for it could rip the bottom planking out of her.

She began to leak, and the leaks increased. What now? Man the pumps. Keep the water down. Sit there and wait, the mariners and the gentlemen (all of whom had worked the long night and the long day through marvellously together at maximum effort and maximum efficiency), stretched on the deck among the cables and the sent-down rigging, snatching a little rest while they could. Wait for the tide to come in again, hoping that again the night tide might be higher than the day's to lift her free her, float her off. The weather remained fine, the sea smooth. The tide rose. But the leak increased: two pumps must be manned constantly, all hands taking turns. Soon three pumps failed to hold the increase. The ship had a fourth but it could not be made to work.

This was desperate. It began to look, if she did come off, that she would sink – go down in the deeper water off the reef whence she could not be salved. What then? Cook had hoped at least to save her, run her ashore, build another vessel from her bones if the hull was irreparable. If she sank they had the boats, but her stores would go down with her: the boats could not accommodate everybody.

At the top of high water she floated, lifting herself like a fat old seal slipping off a rock: but there was now almost four feet of water in the hold. Barely afloat, she stood there in the flat sea above the reef: at least while she was in the same place she could not wholly sink, for the reef would hold her up. For how long? Any breeze would set her grinding, bumping, smashing herself. That razor-toothed coral was no resting place for ships. What to do? Heave off and take the chances? The pumps were holding the water. Only a few men could work the pump-handles: the others were free.

Cook's decision was made. Heave her off! At this moment, as she began slowly to move towards the stream anchor, they had an awful scare. The man sounding in the hold suddenly shouted a depth a foot and a half higher – five and a half feet in the hold! They held their breaths. It *must* be wrong! A glance overside showed Cook that the ship had not settled any further in the water. Check the sounding! A fresh seamen had made it. He had just relieved a mate. It turned out that he had not noted the manner of his measuring. The first took the depth of water inboard from the ceiling – the floor of the hold – the other from the ship's real bottom, and this accounted for the foot and a half difference.

The mental relief of the corrected sounding (though there was still the same water inside the ship as before) 'acted upon every man like a charm'. They redoubled their efforts at pumps, capstan, windlass. She slowly moved out from the vicinity of the reef. She did not sink though the water was now lapping round the chain-plates. They picked up the assorted anchors except one she lay to, for the moment, and a small bower which was lost. They took up the masts and spars from overside, rigged them, got the sails aloft again, set them. The leaking though still considerable seemed under control: but the perfect conditions might not last for ever. Cook headed for a harbour the Master had found by search along-shore, gently under sail, to beach the ship there and repair her somehow. In case the leak or leaks might gain again they 'fothered' her. 'Fothering' was the passing of a sail along the ship's bottom, spread out and littered with bits of oakum, rope's ends, wool and dung from the ship's livestock. The sail was manoeuvred into position below the leak by lines, with the idea that the canvas might lessen the leaks and the assorted flotsam be sucked into the open seams and holes broken by the coral. This was more difficult to do than it may sound, but fortunately Surgeon Monkhouse's midshipman brother had been in a ship wrecked off Virginia and fothered successfully. Cook handed supervision to of the manoeuvre to him and he 'executed it very much to my satisfaction'.

The *Endeavour* was saved. How near a thing it was they saw when she was beached. A large piece of coral broken from the reef was found jammed in the worst hole.

It took many weeks to make the hull sound again. In the meantime the crew camped by the ship, near the mouth of the Endeavour River. Here they really had time and opportunity to learn something about tropical Australia. They ate kangaroo, birds, turtles, clams, fish, the usual greenery including some native beans and stuff Cook called Indian kale. Everything was issued in equal shares, the same for the loblolly boy as for the captain. But there was not an abundance of food. The local aborigines cultivated nothing. Like all their kind they were gatherers only, and they resented the threat of other gatherers on their tribal foreshore and land. Early in August 1770, when at last she was ready to sail again, there were only three months' provisions left aboard – sufficient to reach the Indies but not if she struck upon other reefs. Cook was unable to build up stocks as he had been able to do in Tahiti and parts of New Zealand. The aborigines set fire to the grass round the camp to drive the English away.

There were plenty of reefs still about. From the Endeavour River on to the north was one of the worst places in the whole Barrier area. Cook got out of the lot, heart in mouth all too often, meaning to try for a better passage towards Torres Straits if he could find one. As for Torres Straits, he hoped a useable passage would be found there. One of Dalrymple's maps showed one, for it had Torres's track south of New Guinea eastwards indicated clearly. But Dalrymple's (and countless other) maps showed many things. What was the truth? Cook did not want to find out with his ship's keel. So he tried to get away, beyond the reefs.

But the passage outside was worse – far worse. The trade wind freshened and blew home upon the reefs, which extended in all directions. Seams imperfectly repaired opened up when the ship worked in the seaway. The occasional calms were

ENDEAVOUR

dangerous too, for then the ship had no forward way, while the scend of the sea and the set shoved her steadily back among the reefs again. In calm, where there was not bottom at anchoring depth, he could do nothing. The combined pulling power of sixty men in all the boats would not hold the ship against the current. Arm-power against the sea's set would not do. These reefs rose steeply and suddenly high from the sea bed so that a ship might scrape their windward sides with no hope of keeping off them ... unless the backwash (if any) from the breakers kept her a few fathoms away, and an off-shore wind sprang up.

And so inevitably it happened. On a sunny day in mid

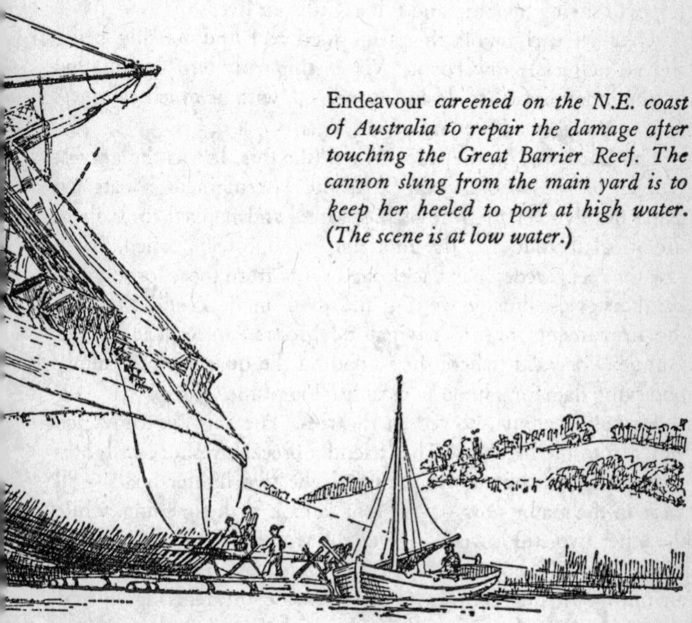

Endeavour *careened on the N.E. coast of Australia to repair the damage after touching the Great Barrier Reef. The cannon slung from the main yard is to keep her heeled to port at high water. (The scene is at low water.)*

August, groping towards the north and hoping soon for an end of reefs for ever, the wind dropped away leaving the *Endeavour* within a mile of 'vast foaming breakers' on a long line of reef, towards which she 'was carried by the waves surprisingly fast'. There was not an air of wind. Anchoring was impossible. Long-boat and yawl were immediately lowered to pull ahead. Cook got out sweeps and rigged them through the leeward ports – the carpenter hurriedly contrived the sweeps from stuns'l booms and planks.

Meanwhile the ship still drifted bodily reefwards: the best that sweeps and boats could do was to turn her head a little away from them. In a little while, says Cook, 'the same sea that washed the sides of the Ship rose in a breaker prodigiously high the very next time it did rise so that between us and distruction was only a dismal Vally the breadth of one wave, and even now no bottom could be felt with 120 fathoms ... we had hardly any hopes of saving the Ship and full as little our lives'.

They were right off the steep-sided reef and steadily being shoved helplessly towards it. 'Yet in this truly terrible situation not one man ceased to do his utmost ... with as much calmness as if no danger had been near.'

The Lord was on the side of men like this. It was the gravest danger of the whole voyage. At the last moment a catspaw came flukily over the face of the waters and flapped the sails to life – briefly but, for the moment, enough. She sailed. They saw the reef recede. The backlashed spray from those 'prodigious breakers' was already wetting the men on deck and clouding the instruments of the navigators, Messrs Green, Clerk, and Gunner Forwood, where they stood on the quarter deck calmly observing data for a lunar to establish longitude.

A small opening showed in the reef. The ship was now 200 yards from the breakers. The 'friendly breeze' freshened slightly – very slightly, but it was enough. The two lighter boats – all were in the sea by now – were sent to look at the opening, while the other two still towed. A great ebb was 'gushing out of it like a mill stream', but Cook put the *Endeavour* in this and got an offing. In due course, he sailed inside the reefs again and stayed there, groping northwards day after day, lying-to or anchored by night, conning from aloft, examining by boats ahead, warily advancing mile by mile through all this treacherous maze to the Straits he hoped to find beyond Australia.

The strait was there. Though it also was a reef-strewn maze, after the Great Barrier Reef its negotiation was almost child's play. The passage had been nightmarish, heart-in-mouth, nerve-wracking stuff for months. Cook sailed on towards Batavia in Java and thanked God there was only one Great Barrier Reef.

In his two subsequent long Pacific voyages he never approached that area again, nor touched at Australia: navigation in the Barrier Reef area could strain a man past bearing.

CHAPTER NINE

Mr Banks is Back from his Voyage

At Batavia, after the best part of two years without mail or news of any kind, Cook learned that the American colonists for whose freedom from French-Indian raiders he thought he had fought in Canada, were refusing to pay taxes of any kind and were, apparently, working themselves into quite a state about the matter. It was 10 October 1770 when he reached Java. The American Revolution was already simmering. Be that as it might, he sent off his reports to Admiralty in London on an apologetic note, for there was no great tidings of a rich and exploitable new continent to give them in exchange for the America they were shortly to lose. Tahiti was Wallis's discovery: Wallis had also seen some of the Society Islands: Tasman had seen New Zealand: the Dutch had already given New Holland everything but an east coast; but he had done his best, and his latitudes and longitudes, he felt confident, were good. The officers and the whole crew had 'gone through the fatigues and dangers of the voyage with that chearfulness and alertness that will always do honour to British seamen'. He added with pride that 'not one man had been lost by sickness during the whole voyage'.

'Sickness' to him meant scurvy, of course – the great avoidable scourge of shipmen. Buchan had died of epilepsy at Tahiti, Sutherland of tuberculosis at Botany Bay, Bos'n's Mate Reading of an excess of rum at sea, Banks' two coloured servants in the cold near Cape Horn, and three men had been drowned. The epilepsy and the tuberculosis were brought aboard with their victims: the other accidents were occupational hazards. There had been the odd touch of scurvy despite Cook's ceaseless and merciless care. Banks had more than a touch – swollen gums, ulcerating 'pimples' in the mouth – so he 'flew to the lemon juice' in his private stocks and the 'effect was surprising'.

Cook's letters to Admiralty were ready when he reached Batavia and went directly back to Europe by Dutch East Indiaman. Never again could he report such a clean bill of health. The ship had to spend three months in Java, rigging down, careening, repairing the hull (further damage was found from that Barrier Reef grounding) and rigging the ship again. Batavia then was a pestilential place of stinking, open drains, an unhealthy port where too many seamen died. Malaria and the 'bloody flux' (dysentery) struck aboard and struck hard. Surgeon Monkhouse was first to die. Tupia and his serving lad soon followed. Banks and Solander very nearly followed them but were moved ashore to an airy house out of the port area and open to the sea breezes. Here Cook sent his personal servant to help to nurse them (for all the Banks retinue were also ill) and stayed with the ship to fight his own fever alone. Several others died at Batavia – seven in all. Forty more were ill and the whole surviving crew (except that wonderful octogenarian the drunken sailmaker) was weakly.

An efficient ship-master could fight scurvy by good discipline, ventilation, cleanliness, reasonable diet and common-sense, but the malignant, murderous bacteria of the Dutch-Indonesian port were utterly defiant. They came hungry and murderous from the malevolent shore and there was no stopping them. In due course, the little *Endeavour* staggered seawards again, her hostages in the cemetery, herself a hospital ship. She sailed into the Indian Ocean by way of the Straits of Sunda and on across by the trade wind belt – that beaten track of square-rigged ships which became the tea clipper road from China – south of Mauritius and Madagascar, into the Agulhas current towards the Cape, leaving her dead sewn in their canvas shrouds behind her as she wandered slowly on. It was a seventy-day passage, fortunately of good weather the whole way – the flying-fish sailors' way of nineteenth-century chanteys. At times the fit men aboard were down to twelve. Astronomer Green, artist Parkinson, Midshipman Monkhouse – the surgeon's brother and Cook's favourite in the young gentlemen's berth – the one-armed cook who had done so well, ten sailors, three of the marines, and even the tough old sailmaker, died. Surgeon

Monkhouse's Mate Mr Perry did his best, but he had no adequate medicines.

More died at the Cape, despite the good conditions there. Cook shipped nineteen men at Batavia to help. Four of these died, too. He shipped another ten at the Cape, several of them Scandinavian and Dutch seamen with names mutilated on the muster roll like Knut Olafsen, Klaus Fick, Johan Bode and Jacobus van Cant. Even after leaving South Africa the deaths continued, though no longer wholly attributable to flux or fever. Robert Molyneux the Master, hailed by the Tahitians as 'Bobba' from his visit there in the *Dolphin*, possibly hastened his end by what Cook calls his 'extravecancy and intemperance' though what this good seaman was extravagant about is not recorded – most likely rum. The ship was in the North Atlantic when Lieutenant Zachary (or Zachariah) Hicks, the competent Number One, died 'of a Consumption which he was not free from when we sailed from England so that ... he hath been dieing ever since, tho he held out tollerable well untill we got to Batavia'.

These were the last to die at sea.

The passage from the Cape of Good Hope to the Channel was one of the most pleasant and simplest for the ocean-going square-rigged ship to make, especially if she came up to Europe in summer. With luck, southerlies from the Cape will bring a ship to the S.E. trades of the South Atlantic: then she is practically home. She must work through the Doldrums, but this is small price to pay for weeks of trades. The N.E. trades, fresh on her starboard beam, will bring her near the Azores, whence it is summer breezes to the Chops of the Channel and home – flying-fish stuff, all of it, ideal for the recuperation of run-down crews and attention to the maintenance of ships. With her people brought to complement again from the Cape the old cat had men enough to eat her. Boats were scraped, cleaned, painted, polished until they were fit for admiral's barges. Masts and spars were scraped, varnished, the yardarms painted; blocks overhauled; the well-used sails repaired and repaired again, running rigging and standing rigging brought as near perfection as possible; the remnants of the livestock scraped from the foredeck though a few sheep from the Cape survived there a while longer and the old goat from the *Dolphin* still prospered; the hempen cables overhauled, and the ship alow and aloft brought to standard for admiral's inspection.

That done, the hands were put to exercising the 'Great Guns and Small Arms', while never once neglecting the compulsory daily intake of 'anti-scorbuticks'. A wan Banks and a paler Solander sat in the shade of the sails on canvas seats aft slowly recovering from assorted fevers, still noting everything – birds, fish, floating jelly-fish, St Helena (a brief call) and the island of 'Ascention' (both on their way) and some shipping including a homeward-bound Portuguese from Rio and a 'Schooner from Road Island out upon the whale fishery'.

There was time for a bit of a 'gam' with the whaler. They visited aboard, bought four large albacore, learned that there was peace in Europe 'but King George behaving very ill for some time but they had brought him to terms at last'. Cook records that 'the America disputes were made up: to confirm this the Master said that the Coat on his back was made in Old England'.* Later the same day they spoke another Yankee whaler out of Boston, five weeks at sea and no whales at all.

Winds were good but not too fresh, which was as well. Both t'gallant-sails blew out so badly they had to be sent down on deck for repairs, and the yards temporarily left bare. The fore topmast backstays parted and the topmast was sprung, forcing Cook to remark that the 'Rigging and sails are now so bad that something or other is giving way every day'. But she sailed along well – 855 miles in six days (336 of them in two) before St Helena in the S.E. trade, 390 in three days in the N.E. A Liverpool brig bound for Oporto spoken off the Scillies seemed surprised to see them. A hairy figure bawled across from the brig that wagers were being held in England that the ship was lost, as no news had been received of her safety for so long.

Then a south-west gale picked up the world-wandering little ship in its noisy arms, and flung her along in a flurry of spray

* See Banks's and Cook's Journals: entries for 18 June 1771. What was 'made up' was one of the disputes about duties, taxation, etc.: in this case probably duties on American imports of English cloth.

and spume, Cook watching his sails and rigging anxiously and the helmsmen their course while Mr Banks's poor greyhound bitch Lady, having survived so much, died quietly on her favourite stool in his cabin. So they passed within a few miles of Beachy Head and sailed on past Dungeness and Dover, while all the Channel shipping stared at the battered old Whitby collier-cat bounding on proudly with pennants flying and the great ensign of the King's Navy stiff in the wind like a ceremonial sail above her quarter-deck. What ship was this? Not Mr Banks's *Endeavour*, surely? For she was reported missing, probably lost. But there was no other King's ship like this.

Good God, it *was* the *Endeavour*!

She anchored in the Downs. It was Saturday, 13 July 1771. There were fifty-six men and boys aboard of the ninety-four who had left England almost three years earlier – fifty-six men and, of course, the indestructible old goat from the *Dolphin*, the first goat in history to survive two circumnavigations.

The arrival in the Downs of the *Endeavour*, 'Captain Cooke, from the East Indies' was duly noted in the *London Evening Post* of 15 July 1771. The 'East Indiaman' had sailed in August, 1768, the sheet recalled, 'with Mr Banks, Dr Solander, Mr Green and other ingenious gentlemen on board for the South Seas'.

They had since made a voyage round the world and 'touched at every coast and island where it was possible to get ashore' etc.

This was about the only mention that Captain 'Cooke' was to get. The same paper a week later disclosed that Mr Banks had discovered 'a Southern Continent in the latitude of the Dutch Spice Island ... Mr Banks passed some months amongst these hospitable, ingenious and civil' new continentals and found them 'politely civilized'. More ships, it was forecast, would soon be sent 'in search of this new terrestrial acquisition'. The *Public Advertiser* of 27 July reported that the crew of H.M.S. *Endeavour* were busily landing 'some of the richest goods of the East' duty-free, as reward for the voyage which, it pointed out, had killed seventy of them, all 'fine, picked men'. Who was left to land the silks, spices etc., it forbore to say.

How the ingenious Mr Green had managed to die was, however, known and announced in an exclusive (from our waterfront reporter?) in the *General Evening Post* the same day. He 'had been ill some time and was directed by the surgeon to keep himself warm, but in a fit of phrensy got up in the night and put his legs out of the portholes, which was the occasion of his death'. Poor Green. Man-eating sharks, one assumes, must have torn his legs off or dragged him, phrensy and all, bodily into the sea.

Mr Banks, 'one of the gentlemen who went to the South Seas to discover the transit of Venus' was presented to George III at St James's early in August, and 'was received very graciously'. 'No less than 17,000 plants of a kind never before seen in this kingdom have been brought over by Mr Banks. ... On Saturday, Dr Solander and Mr Banks ... attended at Richmond and had the honour of a conference with his Majesty on the discoveries they made on their late voyage.' So the astonishing story continued almost as if Mr Banks had retained a personal press agent. A 'Piece of Rock sticking in her Bottom' (the *Endeavour*'s) had saved Mr Banks and his worthy fellow-travellers. 'Very great Expectations are formed from the Discoveries of Dr Solander and Mr Banks, and 'tis expected that the Territories of Great Britain will be widely extended in consequence ...' announced the *Public Advertiser* happily.

Mr Banks was indeed doing well. By the end of August the *Westminster Journal* learned that he and Solander had the 'honour of frequently waiting on his Majesty at Richmond, who is in a course of examining their whole collection of drawings of plants and views of the country ...' What country? Maybe the *Journal* meant the Philippines, whence the highly ingenious Messrs Banks and Solander had been reported (by a rival paper) to have brought 'three different species of the bay, or laurel, tree, from which camphor is extracted by a very peculiar process ...' They had learned the process, too, though it was news to them that they had been to the Philippines. It was soon announced that Mr Banks was 'to have two ships from government to pursue his discoveries in the South Seas, and will sail upon his second voyage next March'. Only two ships for the 'celebrated Mr Banks'? It was three by the end of August, 'with men, arms, and provisions, in order to plant and settle a colony' at Tahiti.

In the middle of all this always ambiguous, often wrong, never corrected stuff – well up to the standards of maritime reporting still generally maintained – a quiet junior lieutenant also apparently back from the South Seas etc. was presented by Lord Sandwich to the King, having waited his turn. The name was Cook. His distinction, apparently, was that he, too, had sailed 'round the Globe with Messrs. Solander, Banks etc.'. A copy of the journal he had kept was presented to his Majesty who, perhaps in return, was reported to have handed over a 'Captain's Commission'.

This was false news, too. Lieut. Cook, at forty-three, might have hoped for promotion to captain after that unique and epochal voyage so splendidly made. But in fact he was stepped up only to commander, which was regarded then as little advance at all. A lieutenant to get anywhere in the Navy had to make captain – the generally accepted next step in those days – and promotion to the rank of commander was considered an assurance of *not* making the rank of captain, or at least of considerable delay in reaching that status. The word 'commander' dated from the late seventeenth-century days of 'Master and Commander', designating a qualified Master who also commanded one of the smaller ships. He was qualified as a Master: as Commander, he could command a ship thought neither big nor important enough for a captain. There were no Lieutenant-Commanders as such until 1914.*

So there was still some air of – well, if not disapproval, at least failure to appreciate the full worth of the man, in withholding the logical next step. The glory, after all, was for Mr Banks who, as one of the eighteenth-century landed gentlemen of England, really had done something outstanding in making the voyage at all. Mr Banks could have his glory, an ephemeral thing at best. Commander Cook preferred his home in the Mile End Road where, after his three years' absence, he learned of the deaths of his little daughter Elizabeth three months earlier, and his son Joseph in infancy. He had never seen poor Joseph, who was born within days of the *Endeavour*'s sailing. Two other sons, Nathaniel and James, were lively little fellows then about eight and seven years old. What their mother and they thought of their father's homecoming – what, indeed, the retiring Mrs Cook thought about anything or did about most things, we don't know. A dutiful wife did not then make a 'fuss' of her husband, share his career or any other nonsense of that sort. She looked after the offspring and ran his home. Mrs Cook did these things well, for being wife to a deepsea seaman was then still a career which worthy women understood and quietly pursued. Cook regarded his private life as his own business. His was not the sort of family which started archives, or kept papers. Cook's writings – like his whole efforts – were his country's. Nor was Mile End a fashionable neighbourhood, visited by the distinguished and the celebrated to mix and gossip briefly and leave long diaries and the like afterwards. Cook was not at all the kind of man the idle or the industrious might chat about, anyway. Even the gad-about Boswell, who met Cook several times, gathered no 'chitchat' from the encounters: or perhaps, like Johnson and so many others of his countrymen, he was aware of his abysmal ignorance of all things maritime and never could get on terms with the extra-

* See Professor Michael Lewis's *England's Sea Officers, the Story of the Naval Profession* (London, George Allen & Unwin, 1939).

ordinary seaman. Dr Johnson's efforts were confined to the production of an epigram* for the famous circumnavigating goat, which was honourably retired to a rich pasture – a better fate than that of many of the old seamen. According to Boswell, the learned doctor had no time for circumnavigations or for books about them either – certainly not Hawkesworth's verbose and flatulent volumes.

Cook consoled his wife – one wishes we did know something about her: she suffered much – and got on with his charts and surveys and other paper work. He was relieved of any nonsense about writing a book, for the celebrated Dr Hawkesworth was commissioned (by Admiralty) to take his rough efforts in hand and convert them into something really readable – some of it mincing, paltry, puffed-up stuff, fit for the eighteenth century. In this respect Banks shared the murder, for his journal was handed over to the prosy Hawkesworth too. So was £6,000, on account, an extremely good advance to any writer in these days and more than the total sum spent on the conversion of the *Endeavour*. Hawkesworth, it seems, was fashionable, a competent hack who could be relied upon to distort or remove all meaning in his personal brand of pompous, bowdlerized prose. There was some readable stuff to be taken from both Cook and Banks.

It must be added in fairness that Hawkesworth, after all, had not sought the job: it was not his idea that Cook and Banks required an editor. The book, in due course, was a huge success in its sales anyway, for it sold several large printings (for those days) at the high price of three guineas, despite the fact that it had been beaten by an unauthorized account, rushed out anonymously by another publisher. This was thought (by Cook and

* '*Perpetua ambita bis terra lactis*
Haec habet altrici Capra secunda Jovis.'

This was made into English by Boswell as follows:

'In face scarce second to the nurse of Jove
This Goat who twice the world has traversed round
Deserving both her master's care and love,
Ease and perpetual pasture now has found.'

One feels such a distinguished goat deserved perhaps brighter lines.

others, who were in a position to know) to be the work of the New York-born midshipman Magra, alias Matra. This Magra, said by some to be originally Corsican, was a bright young man, though not commended by his captain. Perhaps Cook took a poor view of most naval midshipmen, as compared with Whitby apprentices.

Cook was at sea again before either book came out. He was surprised and slightly annoyed later when he read some of the stuff that Hawkesworth had put in his mouth, for Hawkesworth's writing was presented as Cook's first-hand observation. Hawkesworth knew nothing of the South Seas or any other seas, ships or seamen. It was an odd choice. But in the 1770s, who else was there? Alexander Dalrymple?

Alexander Dalrymple, indeed, was an angry, almost a savage man. He found no delight at all in the results of the *Endeavour* voyage, nothing wonderful in the new Pacific maps which Cook brought home. His *Terra Australis* was no longer just *Incognita* but allegedly not there at all. As he had foreseen, the upstart selected for command of his expedition had not even properly looked. The fellow said so himself in his own journal! (Mr Dalrymple, and other privileged persons, had been able to read hand-written copies of this.) When sailing close by land sighted by Quiros, seeing signs of land, he *had not investigated* but sailed calmly along. He would not, he said, 'waste time looking for what he was not sure to find'. Here was not just a dullard but an obvious opponent of the whole conception of the Pacific continent. As for the 'discoveries' claimed for the voyage, these were already indicated in his own *Historical Collection of the Several Voyages and Discoveries in the South Pacific Ocean*, available before the voyage and known to be aboard. Torres' Strait, with the Portuguese navigator's track through it, the east coast of New Holland (very vaguely, it was true, and in one map shown as little but a line drawn from near Cape York to Tasman's Van Diemen's Land, with Quiros's *Austrialia de Espiritu Santo* carried far westwards to become contiguous), and an indication of the Great Barrier Reef – all these were shown. What was Cook's 'Botany Bay' but a poor translation of the *Coste des Herbiages* of the Rotz map, which he must have seen?

These were mean criticisms, even from Dalrymple. As for Torres Straits, he had come upon an account of Torres's passage there after the capture of Manila in 1762, and there is an assumption that Dalrymple at first suppressed this, intending to rediscover the Straits himself, on the way proudly homewards from his discovery of *Terra Australis*. That Rotz map of 1542 certainly was oddly reminiscent of the east coast of Australia as Cook found it and could hardly have been all guesswork.* There were many 'maps', many legends. Who knew anything for truth until Cook came? Dalrymple had long shown himself so much the biased and uncritical collector of hints, rumours, and vaguely charted 'sightings' of cloudbanks, squalls, indications of land etc. that Cook, and others, might well regard him not as advocate but hostile witness.

Dalrymple demanded that another Pacific voyage be made, not knowing that Cook – well aware of the gaps – had already suggested this. It was obvious that as no one had yet made much examination of the South Pacific in high latitudes nor ever a west-to-east circumnavigation of the world down there, there very possibly remained a good deal to discover. Equally obviously, land in high southern latitudes, large or small, was

*Rotz was a French pilot in Henry VIII's fleet who had, apparently, studied maps and charts, and copied his map from a Portuguese original, which is at present unknown. This makes the Portuguese the discoverers of Australia, not the Dutch. They could have been. I have discussed this often with Portuguese scholars. The late Dr Joaquim Bensaude thought it was at least highly probable that after Affonso d'Albuquerque extended the Portuguese empire to Malacca investigators were sent to sail among the islands eastwards searching (among other things) for a mainland or great island said to be there. They could reach Cape York and the N.S.W. coast by small craft, as the Hollanders got as far as the western end of Torres Straits later. The Rotz map was in fact not included in Dalrymple's works given to Banks aboard the *Endeavour*. The Vaugondy chart from de Brosse's *Histoires des Voyages aux Terres Australes* was. This simply made Quiros's *Australia del Espiritu Santo* part of the coast of Queensland, and joined this up hopefully with Tasmania. It is very unlikely that either Cook or Banks then knew about the mysterious Franco-Portuguese maps at all or, for that matter, that Dalrymple did.

probably of more use to the albatross, the sea-elephant and the penguin than to men. Cook may have thought that the chances of finding the useful large land so beloved of the theorists were dim indeed. A man in a ship sailing those tremendous wastes could hope to see only little of them, a twenty-mile riband in a zone of millions of square miles. The behaviour of the wind patterns, of storms and the sea, properly interpreted, would extend his knowledge: but there was a good case for a further voyage.

Cook suggested the idea that it take the form of a circumnavigation in high latitudes, which would examine the South Atlantic and South Indian oceans as well as the South Pacific. Such a circumnavigation could be made by a square-rigged ship in a few months, if there were no interfering land to get in the way. Indeed many such ships did make this kind of circumnavigation later when blown away from Cape Horn, or finding themselves unable to make westing elsewhere. One recalls for example the George Milne barque *Inverneil* of Aberdeen which left Sydney, New South Wales, bound for Bunbury, West Australia, 2,000 miles to the west'ard to load railway sleepers there for Cape Town, in the winter of 1919. The Master found it impossible to make westing at all and so turned eastwards and ran 14,600 miles almost round the world, in eleven weeks, all in the west winds. This was an average of 190 miles a day: the *Endeavour* and similar ships could average little more than 100, and so might hope to complete a southern circumnavigation in five months. But no one then knew how unimpeded such a passage was. It was Cook's business to find out: this he would do with no hurtling rush just because the winds might be favourable and fresh. He needed the southern summers in the Antarctic, not winters – two of them, at least. So the plan he put forward included winter diversions, swings on a large scale into the South Pacific north towards the bases of Queen Charlotte's Sound in New Zealand and Matavai Bay at Tahiti. Using these bases to refresh, he planned extensive investigation in those areas of the ocean, particularly between 100° W. and 160° W. longitude, which neither he nor anyone else had yet investigated. The back-base would be Cape Town (then

Dutch), the forward-bases in New Zealand and Tahiti for use when winter made sub-Antarctic work impossible.

It was the perfect plan. Indeed, it was odd that no one had suggested it earlier. It was bold – what Cook suggested was a combination of the first Antarctic expedition with the first high-latitudes west-to-east circumnavigation and, almost as a sideline, a voyage which would take in, as well, at least as much of the Pacific as the *Endeavour*'s had done. Admiralty approved, as well it might. There were reasons of politics and diplomacy behind this approval. France was interested, too, in the countries of the southern oceans and had sent good seamen looking for some of them – fearless men like Crozet, Kerguelen, Bouvet, Surville, and Marion du Fresne. The French had the advantages of friendship with Spain, and a useful base at Mauritius (then the Isle de France): it was perhaps fortunate that they showed an odd bias for beating about the gloomy, fog-ridden seas of the far south mainly of the Atlantic and Indian oceans. Jean François de Surville was actually beating somewhere near the North Cape of New Zealand when Cook was there, within a hundred miles of him. Surville was outward bound from Pondicherry, one of the French ports in India, looking more or less privately for what he might find and not very well equipped to find anything. From New Zealand he sailed fruitlessly to Peru and was lost overboard a couple of days before his ship made Callao. The next, Marion du Fresne, was in New Zealand not long after Cook, on his way from Mauritius to Tahiti. What he might have discovered no one can say, for in May 1772 the Maoris of the Bay of Islands ate him. It seemed that he had not the knack of getting on with them any other way.

Sooner or later, however, one of these Frenchmen might discover something important, if it were there, and that would not do at all. Already there was reason to fear the loss of the American colonies: Canada of the eighteenth century was poor substitute for them.

No wonder Cook's plans were not just approved but put in hand forthwith. As the *Endeavour* had proved herself so well, Whitby ships would make the new voyage – not the stalwart *Endeavour* herself as she was needed for a store-carrying voyage to the Falkland Islands, but two of the newest and best ships of her type to be found in the London River. Two ships made sense if one were lost: they could assist each other, cover more ground, be each other's 'life-boats' if need arose. Both could be lost together, of course (as the French navigator La Perouse's were a few years later on a reef in the New Hebrides), but it was hoped to avoid this fate. The *Resolution*, 462 tons, and the 336-ton *Adventure* were chosen, both comparatively new, and work was begun on their conversion at once, the *Resolution* at the Deptford naval yard and the smaller ship at Woolwich.

Everything that Cook had learned or could foresee was put into them – frame and timbers of a 20-tonner each to be built as and when necessary, stores and provisions for two-and-a-half years including even more experimental anti-scorbutics, medals showing the head of George III on one side and the ships on the other especially struck for distribution among worthy natives, warm clothing for the mariners when in the far south, and – at last – a Harrison chronometer. Well, not quite that: an excellent copy of one – as demanded by that sceptical, lunar-biased, fumbling Board of Longitude – made by a Mr Kendall. Harrison's 'astronomical clocks' had proved themselves before this: the Board's last stipulation had been the sensible one that these highly efficient time-pieces must be capable of manufacture by other watchmakers. Here Mr Kendall came in.

Both ships were merchantmen. They had to be converted to carry naval crews, stores, and guns. This meant fitting accommodation for 112 men in the larger ship and for eighty-odd in the smaller, but for seamen there was no problem about this. Everyone concerned approved the selection of the Whitby ships and the manner of their conversion, though they were given no special strengthening against ice. They would have to pick their way among icebergs and keep out of pack-ice, when met.

Everyone approved, save one. Mr Banks, delighted with his reception from his first, made it known that he would like also to make the second voyage: but not, if you please, in a paltry tub from the coal trade out of Whitby. He suggested that a ship-of-the-line, a two- or three-decker, would obviously be better – not at all for safety in the ice (there he might have had a point) but for the better accommodation they offered to Mr Banks and retinue. Cook was adamant. The ships as selected must go, he said, and his old captain Palliser supported him. So did Lord Sandwich. Ships-of-the-line were out of the question. A frigate then, said Banks. No, said Cook, Palliser, and Sandwich. And old East Indiaman? No! declared the knowledgeable trio.

In these days it seems odd that Mr Banks or any other civilian could or should question experienced, responsible naval officers and Lords of Admiralty on their practical decisions. Who was he, after all? Not a minister of state, not a powerful politician of any sort, nor an influential member of the Court – not anyone, indeed, but J. Banks, Esq., a landed gentleman of large estate from Lincolnshire who had been a very useful passenger, as amateur botanist, in the *Endeavour* and, before that, a young gentleman down from Oxford who had been elected F.R.S. too young and for too little. He was also friend of Lord Sandwich, and otherwise well connected. His reception from the first voyage, very obviously, had gone to a head that never had appreciated the worth of James Cook. The perhaps subconscious resentment of his 'ship-driver' – that dour thwarter of shore excursions, that martinet, that enthusiast for anti-scorbutics and Whitby colliers – flared into active life.

Mr Banks was not accustomed to being thwarted. He had, after all, provided most of the scientific staff for the first voyage by his own selection and at his own expense: he was prepared to do the same for the second voyage and was, indeed, well forward with his plans. In all England, he was the *only* man who had devoted his own time and fortune to such enterprises, or was prepared to do such a thing – to hazard his welfare and indeed his life on dangerous and unprecedented chases round the earth in (to his view) inadequate ships. There was a lot to be said for Mr Banks, indeed: he said it.

Since the *Resolution* was already selected and bought, a compromise was reached to alter her by adding considerably to her accommodation. Banks was particularly concerned that the great cabin should be adequate for the large party he was engaging – it included Solander, of course, the artist Zoffany, naturalists, scientists, draughtsmen, servants, and a personal orchestra of three, making a total of fourteen – and the great cabin was the only recreation and working space for all these. Later it came to light that the thoughtful Mr Banks meant to provide yet another, very odd, member of his party. Mindful perhaps of M. de Commerson's valet-wench recognized aboard Bourgainville's frigate at Tahiti and the obvious warmth and value of such companionship in Antarctic weather, he had arranged for a young woman, disguised as a boy, to join the ship unobtrusively later at Madeira – a Miss (or Mrs) Burnett, from Edinburgh – to be his 'valet', too.

The great cabin was also, of course, the only chart-house, navigation headquarters, captain's and senior officers' ward-room (by captain's permission). So the planned great cabin must have its deckhead raised for better head-room, and be extended forward. Since even with all this, Cook would be crowded out of it, an extra cabin was to be built above the lot for him – a sort of Master's doghouse such as seventeenth-century ships often carried. The maindeck itself was raised and another house built on that. These alterations were considerable and expensive. It was also immediately obvious to any seaman that they must affect the working of the ship – indeed her very stability and seaworthiness.

Ships of the seventeenth century and earlier were greatly built-up, as all the old pictures show: but their superstructures were lightly built additions hammered on by house-builders after the ship herself had been solidly constructed to her main deck. The additions to the *Resolution* were very different, for they had to be shipwrights' jobs, heavy, solidly built additions of the stoutest construction, fit for the Roaring Forties, Cape Horn, and the Antarctic. The over-built decks affected the leads of the essential gear to work the sails, interfered with the set of the courses, and generally cluttered up the ship. Cook is alleged to have accepted all this. He must have been appalled. Spoil a good ship? It could only be absolute anathema to him. But he watched the procession of the high and mighty, the exalted and the nincompoops who came to Deptford almost daily to visit the ship 'in which Mr Banks was to sail round the world'. He bided his time. Her real condition would show up very clearly as soon as she came out of dock.

It did. She was useless – unstable, crank, a floundering death-trap, rolling violently from side to side even in the flat water of the Thames whenever a few men crossed the decks. The Banks servants and musicians, drawn up on deck in special uniforms of scarlet and gold, had to be careful of their footwork to remain upright. The ship was drawing seventeen feet of water, too, instead of the fifteen feet designed: there was no easy answer to her lack of stability by adding more ballast. Indeed, they must take out a lot, which was already built up in the hold.

'I cannot answer for her', declared the experienced old Trinity House Pilot, making his voice heard above the Banks band. 'I won't work her, not even to the Downs. She'll fall over.'

One suspects that Cook knew that, long before. Now, everybody could see. It would be very odd if he did not, if he and all the master shipwrights were so ignorant of the stability problem in so long-established a ship-type as the Whitby collier that they had no idea what they were doing. But the instability was not just in the ship. Had they decided that there was only one way for the ignorant and arrogant to learn? As a seaman with some odd experience of a similar situation, one puts forward the idea. The ill-treated ship, having made her impressive protest, had to be hauled quickly into the nearest dock, which was Sheerness. What now? The only answer was to rip all the expensive new houses, decks etc. out of her, or she would never be fit for sea at all. Mr Banks, his scientists, friends, servants, and musicians would have to accept her as she was.

Mr Banks was as publicly committed to his voyage as Mr Dalrymple to his continent. Now he was furious. He came to Sheerness and 'swore and stamped upon the warfe like a Mad Man', ordered everything and everybody of his out of the ship, went home to write arrogant letters to Sandwich like a man who had learned nothing. There was still no end to his demands. A West Indiaman would do him now – any ship but the

Resolution, under a captain whose skill might cope with ships not colliers. This was a nasty swipe at Cook. There was a paper war, a campaign among the mighty and in the House of Commons, accompanied by some exposure of things not recorded in the Journals. It appeared that the Banks-Solander party had been highly critical of the *Endeavour*, too (which is understandable and acceptable: after all, they had been jammed aboard for the best part of three years: the pity was that Banks now mentioned it), and obviously resented Cook's calm inattention to the desires of his passengers when the ship's needs dictated otherwise.

Mr Banks writes very oddly indeed, in the strain of a petulant spoiled boy. There was no longer any doubt that Cook's second voyage was to be Banks's, and the command thereof.

'We have undertaken to approach as near the Southern Pole as *we* can' – the 'we' being really the first person singular. 'We are to attempt to pass round the Globe through seas of which we know no circumstance ...' and so on, all written as if the complainant was the sole organizer of the voyage and the ships were being 'fitted out wholly for his use, the whole undertaking to depend on him and his people, and himself as the Director and Conductor ...' as the Navy Board commented in a minute on one of these hot-headed letters.

Banks did not desist. He just could not accept the fact that he could be overridden. Not even a straight broadside from Palliser – annoyed at the aspersions on Cook – and another equally to the point from his old friend Sandwich had any effect, for the time being. Indeed, the Banks statements, demands, and arguments grew wilder. It was obvious that, in addition to his criticisms of the ship, there was to be no nonsense about Cook being in command. That came into the open at last. Cook was to do what Banks told him. When this preposterous demand was turned down, the next was that he should at least control the officers' preferment – that is, promote them or not as he chose, after the voyage. This was so obviously a mere variant of the previous notion and indefensible attempt at interference in the naval service that silence was the only answer. Banks sulked away.

It was all rather childish and stupid, too. After all, if he insisted on dictatorial command, Banks was in a position to charter or buy his own ship and do what he liked with her, though perhaps not a ship-of-the-line. Captains Carteret and Wallis were still about, and Carteret available, for that ill-treated officer had not been employed since the *Swallow* returned. This solution appears to have occurred to no one: perhaps the embittered Carteret was a hard man to get along with, and Banks was at least aware that he knew the best ship-driver.

In due course, the *Resolution* and her consort sailed, with replacement naturalist, artist and draughtsmen. The diverting prospect of a small orchestra, in scarlet and gold uniforms, offering concerts for penguins, seals, and skua gulls listening from icefloes, while the valet Burnett dallied prettily in the cabin, faded for ever.

For Mr Banks found himself back from his voyage indeed.

CHAPTER TEN

Icy Circumnavigation

THE *Resolution*, Cook's 'flagship' for this second world voyage, was not just a large copy of the *Endeavour*. Speaking technically, she was a ship-rigged sloop-of-war, neither bark nor barque. According to Falconer's *Universal Dictionary of the Marine* of the time, sloop-of-war was 'a name given to the smallest vessels of war, except cutters. They are either rigged as ships or snows.' A snow was a brig with an additional lighter lower-mast stepped immediately abaft, and as part of, the mainmast itself: she was two-masted, of course, or she was not a brig. The Holman painting, a good professional job of conventional ship-portraiture of the period, shows the *Resolution* as a full-rigged ship of good profile and handsome seat in the water (unlike the snub-nosed, homely old cat), rigged with three standard square-rigged masts, the mizzen topgallant-mast stepped at the mizzen topmasthead. In other words, the artist gives her three masts rigged in the full-rigged ship manner of lower-masts, topmasts, and topgallant-masts. It is true that some other artists do not show this but give her a mizzen similar to that of the rather ambigously-rigged *Endeavour*. One accepts Holman, the commissioned marine artist, though later in the voyage when spars and lighter masts began to go it is possible that Cook re-rigged her, omitting the fidded mizzen topgallant. It made little difference. Good spars were short in the South Sea islands.

Sloops-of-war, says Falconer, carried from eight to eighteen cannon: the plans of the *Resolution* now at the National Maritime Museum at Greenwich show her pierced for eight guns on the starboard side, and drawings show two stern-chasers. She therefore could have had eighteen guns but carried twelve, which was enough. They were four-pounders. Her complement was 110 men. She had a graceful, sweeping cutwater surmounted by a figurehead. Holman paints this as a sort of outsize animal head, but it is not shown on the sheer draught: if the commissioned Holman painted it, it was there, for figure-heads then were often attached as embellishments after construction, not built into the ship. The protruding cut-water with its associated gratings and sort of platform beneath the bowsprit gave the seamen two advantages – an easier way to the bowsprit and jibboom, and a more comfortable 'heads' than the *Endeavour* had – perhaps one should say, less dangerous: for there could be no real comfort for exposed seamen in that windy spot. It must have been trying at all times and exceedingly so in Antarctic waters.

The stern and counter, too, were more handsome than the *Endeavour*'s, and the great cabin was surprisingly ornate (if one is to believe the voyage artists), a seven-windowed apartment, spacious, well lit, and airy. The decks were roomier, the accommodation, and the space for gear, provisions and stores better and larger. The *Resolution* had been built at Whitby as the *Marquis of Granby* for the coal and short-sea tramping trades. Her burthen was 462 tons, almost 100 tons more than the earlier ship (which in small ships is a lot): her principal dimensions were 110 feet 8 inches length by 35 feet 5½ inches maximum beam by 13 feet 1½ inches depth of hold. She was sold to Admiralty for £4,151 and the conversion bill was another £6,565, mainly wasted on Banks. She had the same flat-floored, apple-cheeked, comfortable hull as the *Endeavour*, built to carry well, sail adequately, and sit upon the ground in safety when necessary. She was not coppered but sheathed with a double skin of wood, with the usual borer 'deterrents' packed between the hull proper and the outer skin. For the Antarctic this was better than copper plates which even brash-ice could rub off: nor was the proper coppering of ships' bottoms understood at the time. Cook had no desire to see his rudder drop off. As a Whitby-man her complement was perhaps twenty, including the usual crowd of juvenile apprentices. Admiralty commissioned her with ninety seamen and eighteen marines, as well as the usual (plus some unusual) supernumeraries.

The *Adventure*, 340 tons, Lieutenant Tobias Furneaux in command, was a smaller copy of the *Resolution*, built at Whitby a year or so earlier as the coastal collier *Marquis of Rockingham* and bought for £2,103 into the Navy especially for the voyage. Cook chose both ships. She was double-skinned too. She was only some thirty tons smaller than the *Endeavour* and went to sea with fifteen fewer persons on board, and nothing like the Banks party to clutter up the great cabin and the quarter-deck. Conversion included the essential accommodation, below-decks galley, sail-lockers, ammunition lockers, cabins for the officers. She was a straightforward little ship, stout and good-looking but – under Tobias Furneaux at any rate – perhaps not a very good sailer. Lieutenant Furneaux was unused to Whitby ships, and he never did seem to get the knack of tacking her. He was an excellent seaman, a Devon man, navy-trained, and he had been second lieutenant of the *Dolphin* under Wallis. Perhaps that was his trouble. Both Wallis and the first lieutenant were ill much of the Pacific voyage and virtual command passed to Furneaux. Aboard Wallis' *Dolphin* the accent was on minimum risk though the frigate was an excellent sailer.

Though both, apparently, had served in the Canadian campaign, Furneaux was a stranger to Cook. So was Cook's own first lieutenant in the *Resolution*, perhaps because the experienced officers of sufficient seniority like Hicks were dead. His name was Robert Palliser Cooper: his middle name suggests the source of his recommendation. He could not have had a better. He had been on the North American station, too, and was destined to make post-captain almost as soon as Cook did, and rear-admiral towards the turn of the century. After the *Endeavour* voyage, though Banks and cohorts had taken – or been given – the credit ashore, career naval officers well knew and respected the name of James Cook. There was no war on at the time, though the War of the American Revolution was coming up: a share in a voyage of discovery with Cook was a good thing to do for those career officers who might survive it and, under Cook, the chances of doing that seemed good. Not for the ordinary mariners, perhaps. Many of these deserted from both the *Resolution* and the *Adventure* before the ships sailed – nearly sixty from Cook's command and another forty-odd from Furneaux. The ships were perhaps too close to London. The death-rate in the *Endeavour* on that sickly homeword run from Batavia had been bad enough though in those days not considered excessive or even high: but a man was as dead from the bloody flux as from scurvy. There was understandable reluctance to accept undue risk of death from either. The coming voyage was known by lower-deck 'buzz' as a voyage into waters no man knew, an icy and interminable waste at the end of the world somewhere which at the very least would be both lengthy and uncomfortable – not that most seamen knew much of comfort, but there were limits. At any rate, they fled from the ships by the score.

Those who stayed were the better off for the loss of the deserters, and Cook saw that they were well treated. One grievance seamen had then (to flare up in the Spithead and Nore mutinies before the century was out) was lack of pay. Their alleged pay was a pittance, but they seldom received even that. Cook's seamen were paid not just all that was due them, before sailing, without the usual exasperating deductions, but two months' advance in cash as well. (This would surely be the time for the disaffected and the mutinous to desert: the ships were still at Plymouth, the land close. But nobody did.) There were twenty *Endeavour* veterans in the *Resolution* including three officers, three midshipmen, the officer commanding the marines and one marine, and thirteen seamen. This was a considerable proportion. The afterguard of the *Resolution* was sprinkled with famous names. George Vancouver was there as a fourteen-year-old midshipman; James Burney, the famous scribbling Fanny's brother and an excellent writer himself, to become later F.R.S., rear-admiral; Joseph Gilbert, Master, another excellent seaman and surveyor whose family name is honoured in the Gilbert Islands of the Melanesian Pacific; Richard Grindall, who as Captain fought the *Prince* at Trafalgar under Nelson, was knighted and became vice-admiral. Cook was already a notable officer with whom to serve, and bright young officers knew it.

The *Resolution*'s crew broke precedent not only in the receipt of full and advance pay: they were given a whole cook, an almost unique salt-horse-boiler, with two arms complete with hands *and* two legs. His name was John Ramsay, and he knew what he was doing. He had been an A.B. in the *Endeavour*. Nor did all colour go out of the expedition with the departure of Mr Banks's liveried orchestra. Perhaps taken with that idea, Cook asked for and got two 'Highland pipers', and a marine drummer who could also play the violin. Other marines could perform with the fife. This lot was hardly for the delight of the gentlemen of the great cabin, for their music might have been ear-bursting there: but the pipers made history in Tahiti and the Islands and the skirl of their bagpipes hurled defiance at the Antarctic winds.

As for stores, the ships were laden down with experimental beer extract, worts, robs, saloups, portable soups and all manner of anti-scorbutics strange and familiar, rum, brandy, gifts, hardware and other goods to trade with the Maoris and Tahitians and all the rest for fresh provisions. Cook's list of provisions survives in a copy of his journal. It includes such items as 59,531 lb. of biscuit, 7,637 four-lb. pieces of salt beef and 14,214 two-lb. pieces of salt pork, 19 tons of beer, 1,397 gallons of spirits, 1,900 lb. of suet (to stop the seamen using cook's slush on their biscuits, a scurvy habit), 210 gallons of 'Oyle Olive', 19 barrels of 'Insspicated Juce of Beer'(condensed extract of beer), nearly 20,000 lb. of 'Sour Krout' and 30 gallons of 'Mermalade of Carrots' as produced and recommended by the Baron Storsch of Berlin. The portable broth, saloups, and salted cabbage were all there, but unfortunately the lemon and orange juice were still just medical stores and there was not nearly enough of either. Both ships carried livestock – a small bullock or two, sheep, goats, hogs, poultry including geese – but the numbers are not given. Bullocks were quickly slaughtered and fodder for sheep could be carried only in small quantities, but the real trouble with livestock was their high consumption of fresh water. Officers were allowed – indeed expected – to bring sea stocks of preserves, pickles and whatever else they fancied of their own, to help out, and there were some cooking facilities in the pantry aft in addition to the rather primitive galley right for'ard. Large supplies of fishing gear were distributed, and galley equipment included three different systems for small-scale distillation of sea-water or purifying foul fresh-water, and a stove for baking bread. This, unfortunately, could be used only if set up ashore, and there was no baker.

Astronomy and art were also provided for. An astronomer named William Wales (an excellent and able Yorkshireman) was there to take care of the copy of Harrison's chronometer and to assist generally with observations. With him came William Hodges, 'landskip painter', signed on as A.B., really the official artist. Most usefully, the crew included at least two men who had experience of Greenland whaling. Whitby was a port for Arctic whalers as well as North Sea tramps and colliers, and sent a considerable fleet to the fisheries each summer. Arctic conditions were very different from Antarctic, but apart from these whalers, no one in either ship knew anything about ice or ice navigation at all.

There was also on board, and most unfortunately a scientist – perhaps one should say an alleged scientist. This was the Scots-Prussian John Reinhold Forster, ex-pastor, ex-schoolmaster, ex-and-future hack, 'a litigious quarrelsome fellow' indeed. If Admiralty had set out deliberately to inflict upon Cook and everybody else in the *Resolution* the most troublesome, useless and energetically hostile a shipmate it was possible to find, they could not have made a better choice. That perceptive chronicler but most mild of men, Dr J. C. Beaglehole of New Zealand, finds even his quiet pen slipping hot with wrath across the page at the very thought of – well, not just Forster for he could be left to his God, but the infliction of so perverse a wretch on the unfortunate and long-suffering Captain Cook over so long, and difficult, a voyage. 'Dogmatic, humourless, suspicious, contentious, censorious, demanding, rheumatic, he was a problem from any angle,' Dr J. C. Beaglehole summarizes. 'From first to last on the voyage he was an incubus. One hesitates, in fact, to lay out his characteristics, lest the portrait should seem simply caricature.'

What on earth a botanist even of the highest skills could find to do in sub-Antarctic and Antarctic waters (which it was the prime purpose of the voyage to search) it is hard to say, especially one who did not and could not land there. Mr Forster had never made a voyage except across the English Channel, had not the slightest interest in seamen or the sea, nor the Antarctic nor Pacific, nor indeed in anything but Mr Forster and the £4,000 which had been put up for a 'man of science' to go on the voyage. He was in fact a very bad penny which turned up after the Banks imbroglio; that £4,000 had been voted for a man of Banks's selection who could have been relied upon for two things, at least, that he would be a gentleman and a good shipmate. But the Banks party had swept off in a 'huff' with their youthful leader on a jaunt to Iceland. The standard of selection by civil service commissioners – if any – was not then as good as it later became. Hearing of the soft berth going, Mr Forster offered and seems able at least to have mustered a little temporary charm to go with the offer: in the absence of anybody else, he was accepted. The sum of £4,000 was a small fortune in the 1770s, enough to set up a prudent man for life. By contrast Cook's naval pay was six shillings a day: the whole three-year voyage was to bring him little more than a tenth part of the princely sum squandered on J. Reinhold Forster at its beginning.

It is fair to add that Forster did make one contribution to the voyage. He brought his son, George, a clever lad – he was seventeen when he joined – who really had some knowledge and considerable ability and could be human too. George in the end was to do a very mean thing though at his father's instigation – he cheated on the book. He rushed into print with an *Account of the Voyage etc.* which did not just misrepresent Cook to the point of libel but was planned to rob him of the royalties on his own official account. This could not be rushed out, because it had to include engravings of the official charts and illustrations. There was an understandable prohibition on prior publication of books by the ship's company: but Forster the elder had spent the £4,000 long before – no one knows how – and was clamouring for more money. So he produced a dishonest book through George, who was not (he said) affected by the prohibition. This time, unlike the first voyage, Cook was to be allowed a share in the book produced from his journals.

Those stalwart and unfailing James Cook supporters, the First Lord, Lord Sandwich, and the Comptroller of the Navy, Sir Hugh Palliser, went out of their way to show their approval of the great seaman and to make certain that he and his men had all they needed by making a special visit aboard before the *Resolution* sailed. This was the same Sandwich who was so ardent a gambler that he did not give himself time to eat, and his habit of hastily slapping a round of beef between slices of bread and wolfing the lot down gave his name to the well-known sandwich. He was a politician, not a seaman, and as such his administration of the Navy was sometimes criticized. But he was aware of the merits of James Cook, and he fostered them.

The *Resolution* slipped out of Plymouth Sound early on the morning of 13 July 1772, with an accompaniment of bellowing bullocks, grunting hogs, cackling fowls and guttural goats and the bewildered sheep scampering for foothold in their pens. All these were customary shipboard sounds, proper at a ship's sailing. Not so the sight (and sound) of Mr J. R. Forster quarrelling loudly with the Master at the for'ard end of the quarter-deck. Mr Forster wanted the Master's cabin which, he declared, was better than his own and had been allotted by favouritism. In naval ships the allocation of cabins is a serious matter done very carefully: once done, it stands. Mr Forster got nowhere with his shouting, though he offered a bribe of £100 for the Master to change. In the end he shouted very foolishly that he would tell King George about the matter when he got home. This gave all hands a great laugh and became one of the ship's best jokes. Seasickness soon drove the argumentative fellow below decks. Sixteen days later the ships were at Madeira: everybody else settled down well. At Madeira the two ships took in the usual stocks of the island's excellent wines, and bought fruit, vegetables, fish, and more livestock for fresh food. The *Resolution*'s vegetables included a thousand bunches of onions for distribution to the crew: this time nobody had to be lashed for refusal to eat the fare provided.

Now, too, Cook learned for the first time of the Banks scheme for providing himself more agreeable company. A 'Mr Burnett, a person ... about 30 years of age and rather ordinary than otherwise' had hurriedly left the island, he wrote, a few days before the *Resolution* arrived, after waiting for Mr Banks's arrival about three months because, he explained, 'he could not be receiv'd on board in England'. 'Mr' Burnett obviously lacked the skill, the figure, and the discreetness of M. de Commerson's Mlle Jeane Baret, for Cook reports that 'every part of Mr Burnett's behavior and every action tended to prove that he was a Woman; I have not met with a person who entertains a doubt of a contrary nature'.

A sight of Palma in the Canaries and a brief call at the Cape Verde Islands for fresh water and more livestock made a break in the long run to the Cape of Good Hope. These Portuguese islands of Madeira and the Cape Verdes lie in the north-east trade wind belt very conveniently on the way. It was necessary for Cook to make towards the Antarctic in the summer. To do this comfortably he had no time to lose. His calls were brief. The ship's routine had long been running smoothly. She was on the three-watch system, with the large crowd of midshipmen joining in. These had first to work as sailors and learn properly to 'hand, reef, and steer' – the classic requirements of the square-rigged able seamen. They 'handed' sail, which meant helping their watchmates aloft to fight the deep and awkward tops'ls in a gale, to set and shift topgallantsails, studding-sails and any other sails: they had to reef too, no matter how hard the wind or violent the weather: and, though it was not part of their duties, they must get the feel of the ship by mastering the art of steering. The grog, beer etc. flowed freely to help things along, and the makings of a good 'flip' were abundant.

The crew fished industriously and caught a few bonita now and again. Scared flying-fish, taking off with difficulty from the roll of foam before the bows, were caught in the backdraught from the sails and crashed on board. These were odd variations in the fixed round of meals: there were rarely enough fish to go round. Cook established the usual sea routine of long voyages with three meatless days a week, called 'banyan' days, when butter and cheese and 'as much boiled pease as they could eat' were dished out to the messes. Each man was entitled to a daily ration of a pound of biscuit, as much small beer as he could

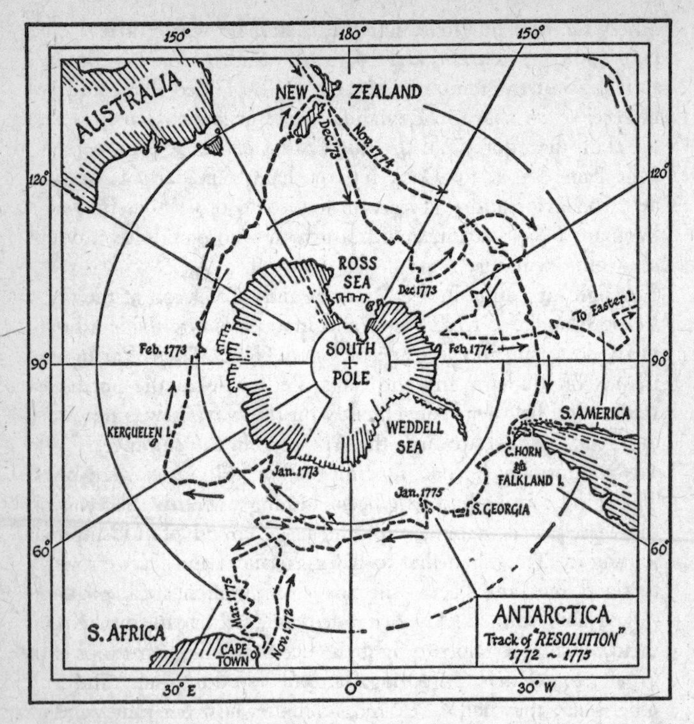

drink (if available) or, in lieu, a pint of wine, or half a pint of rum, brandy or arrack. Sundays and Thursdays were pork days. Slabs of the glue-like portable soup were boiled with the pease, oatmeal, or wheat. Sour 'krout' went with the boiled beef. Cook explained that it was 'cabbage cut small and cured' somehow – he was not sure just how – and 'much used in Germony'.

All this time the *Resolution* sailed along very well, in spite of Banks's fault-finding, with never the slightest hint of instability or any other manner of unseaworthiness – it would be astonishing if there were – and the *Adventure* usually kept up with her. Neither was fast. Days' runs of more than 150 miles were unusual, and anything over one hundred was good. They were 109 days to the Cape, which was not bad for two small ships in company. Cook's instructions directed him to look for a mysterious place called Cape Circumcision, said to have been seen by the Frenchman Bouvet, and possibly part of that ever-missing 'continent'. So he headed south from the Cape, after replenishing fresh food and water, and searched assiduously but in vain. At last, with the opinion that Cape Circumcision was probably the top of an iceberg and in any case no part of any continent or land of any size, he gave up and moved on.

Actually there was a small island in the area, now known as Bouvet Island, which has the distinction of being the most remote and isolated speck of land upon earth. There is not another isle, islet, or piece of rock within a thousand-mile radius: looking for such a place down there was like groping for a pinhead on a fogged-in airport. With the Harrison-Kendall chronometer-watch on board, Cook's navigation was of even higher standard than by lunars; the trouble was that he had to look for the 'Cape' where Bouvet said it was. It wasn't. The weather was often poor, stormy and cloudy, for they were soon down in sub-Antarctic waters. He continued towards the south. What possible use to anyone a cape or continent discovered down there might be no one could say: but the instructions were clear. If Cape Circumcision eluded him, Cook was 'to proceed upon farther Discoveries, either to the Eastward or Westward as your situation may then render most eligible, keeping in as high a Latitude as you can, and prosecuting your discoveries as near to the South Pole as possible ...' This in a couple of unprotected merchant sailing-ships with open decks and not even a wheel-house or shelter on deck of any kind! Or really adequate warmth anywhere.

Ships like the *Resolution* and *Adventure*, as square-rigged ships, could be sailed and worked only from their own exposed decks (where all the sails' gear led, the labyrinth of controlling tackles, cordage and lines with their curious age-old names, their precise places, functions, and manner of working) and in their cruelly exposed rigging. Their sails were planes set and trimmed to catch the wind and convert its force to forward speed: to set them, reef them, 'hand' them it was essential for men to go aloft, often in large numbers. The rigging froze with the moisture coagulated into its fibres: the sails froze into stiffened statues of themselves touched to white marble. It was as easy to 'hand' them as to furl pressed steel, as simple to haul frozen lines through frozen blocks as make handy-billies out of iron-wire shrouds or shift Cleopatra's Needle on the Thames Embankment with a score of telegraph-poles. To touch the frozen rigging on frigid days was to risk frost-burn which sears like flame: to fight those iron-hard sails aloft meant bloody hands and minced fingers, nails torn out by the roots and the hot blood swiftly frozen. This work could not be done with gloves. These men had no gloves – mittens at the wheel, yes: no gloves aloft. Canvas won't yield to gloves, and a sailing-ship sailor must have his 'feel' to do his work. In the intense cold with the frigid breath of frozen hell back-swept from the huge icebergs all round them, the human mechanism found places to freeze not hitherto thought of – the moisture in the eyes and in the nostrils, even among the hairs of the moustache and beard. Prosecute discoveries as near the South Pole as possible? What desk-bound bureaucrat wrote that? Discoveries of what? That cold hell long continued may be worse than burning?

A good place for some bureaucrats, perhaps: but the bureaucrats were not there.

The trouble was that Northern Hemisphere dwellers could not help thinking of the greatly differing Southern Hemisphere in their terms. Many little ships no stronger and no better equipped than the *Resolution* and *Adventure* had been in the Arctic and Arctic waters, in quest of the illusory North-West Passage or the equally hopeless North-East, or after whales. But Arctic and Antarctic waters are quite different. It is not possible to circumnavigate sub-Arctic seas except in a nuclear-powered submarine or massive ice-breaker, for the Arctic in essence is a frozen sea surrounded by great areas of land: the Antarctic is a great frozen land fringed by iceberg-littered pack and surrounded for the whole girth of the world by tempestuous seas, in which the 'berg and growler belt – the ice-line – extends well into the temperate zone. Europe's Arctic whalers worked by summer only and could, they hoped, avoid the pack.

Arctic explorers must pit their ships against the ice – in the end, get caught in it, drift with it for years to get anywhere (as the great Norwegian polar men did). A ship drifting in Arctic ice can hardly be sailed. Unless she is crushed by ice movements or forced up and toppled over – some were – she may survive to get somewhere: in the meantime she may make tolerable quarters. If she be not held in ice by design or accident, she may sail freely back to Europe, not so very far away. Eskimos, Greenlanders and Lapps live freely the year round in the Arctic. Nothing human survives in the Antarctic. The Arctic offers animals in abundance, the succulent salmon, the good-eating duck and countless geese and other birds, the 'king' crab, a thousand growing things. 'The friendly Arctic', Vilhjalmur Stefansson called it, and made a good case for the name. No one has called the Antarctic friendly. The seal, the penguin, the whale, some petrels and the skua gull live there: none is good eating. Even the hardy albatross gives the place a berth. Nothing useful grows beyond the Antarctic Circle: nothing ever will survive that fiendish climate.

All this Cook and his crews had to find out. The weather was not bad, the wind not always howling, the temperature not always well below freezing. Sometimes there were good days of quiet sea and reasonable weather, with fresh water gushing from big growlers and rotting 'bergs. Cook put out boats to gather ice to melt aboard: there was at least no shortage of fresh water. Soon the ships were sailing through a sea littered with floes large and small, growlers – smaller icebergs, largely awash and hence difficult to see and very dangerous: neither ship had as much as an iron-protected cut-water – and icebergs so enormous the sailors called them ice-islands. At first many thought them menacing though beautiful – 'a noble awfull seen', wrote Lieutenant Pickersgill in reluctant admiration – but they became used to them. With care, under easy sail, the ships could survive among them. In January 1773 Cook sailed across the Antarctic circle – the first man ever to do so – not far from Enderby Land. He came within seventy-five miles of land soon afterwards but he saw none. Always too much ice lay between him and the frozen continent, and visibility was often poor.

A great, apparently endless zone of solid pack-ice blocked his path to the south. Not only was the pack impregnable but if a lane did open into it and the ships entered, they could be nipped inside – for ice-lanes close as well as open – and caught.

'We discover'd from the Mast-head 38 islands of ice extending from one bow to the other,' Cook's journal records for 18 January 1773. 'Soon after we discovered Field or Packed Ice in the same direction and had so many loose pieces about the ship that we were obliged to loof (luff) for one and bear up for the other ...' In other words, the ships had to be conned through a maze of nasty big floes, growlers, and icebergs, hauling close to the wind to weather some, falling off to run from others, with the area of sea free to sail in steadily decreasing, and the ice thickening and solidifying towards the south and east. The 'Greenland men' were obviously getting a little concerned about it, as well they might. So was Cook. He had come as far as a sailing man could go. It was no longer 'consistent with the safety of the Sloops or any ways prudent' to stay there: a 'strong gale attended by a large sea' got up from east, bringing poor visibility with sleet and heavy snow: he close-reefed the tops'ls (giving all aloft an extra glass of brandy as they came down from aloft) and head-reached her northward away – he hoped – from the ice.

One likes that glimpse of Cook, that taciturn titan of the South Seas, standing on his quarter-deck supervising the issue of a deep glass of straight stuff to his chaps as with their Fear-nought jackets soaked with the sleet, the red-baize balaclava caps he'd given them jammed on their heads until only blue noses and frozen whiskers showed, their tough hands torn and fingers bloody, they balanced themselves with grace and ease on the iced-up decks and tossed the good liquor back with a grin and a look of quiet defiance at the elements. Brandy never served its purpose better, giving them all – man and boy – a warming 'lift' as they made their ways to their wet and some-what cheerless quarters below. It was morale-building: they could turn in now with warmth, reflecting on a job well done and, perhaps, on the satisfaction of serving a captain who did his obvious best always to look after them. The snugged-down ship could be left to her own devices for a watch or two or until the wind eased: they could sleep. The ships pitched and rolled and the great 'bergs tossed and rolled all round, sometimes grinding together like the heads of fighting bull-elephants, cracking and banging with sounds like cannon-fire, with now and again one capsizing, rolling over in awesome, frightening grandeur, froth-ing the torn sea as if a meteor had suddenly entered it from space.

Many of these 'bergs had wondrous shapes and on fine days could be movingly beautiful: but this was neither time nor place to admire beauty. When the strong gale eased both ships set more sail and made speed to the north. Even Cook did not dare cross the Antarctic Circle again that season. So he searched for more French-reported discoveries such as the large Kerguelen Island and Crozet's group, further to the north, then slipped south again to make easting beyond the heavy ice past the whole of the Indian Ocean Antarctica as far as the present Adelie Land. There was no slightest sign of any land, useless or useful – nothing but the storms of the 'shrieking fifties' as later-day sailors called the area, and intense cold. The ice lessened in area slowly at first, then quickly: first the closed pack disappeared, then the broken floes and growlers – not before the *Resolution* bumped a few of them, fortunately with little way: with summer then so far advanced their jagged points and sharp edges broke off before doing damage – and at last the 'bergs.

In a hard blow with thick weather the *Resolution* lost contact with Furneaux's *Adventure*, which disappeared. It was remarkable that the two ships had been able to keep company so long in such bad conditions, and only Forster senior was alarmed: the crew suggested he tell King George about it, for those rough, swearing, cursing, grog-swozzlers – as he saw the brave, hard-working fellows – refused to take him seriously. They knew that a possible parting was long foreseen and the *Adventure*, if adrift, had her orders to make for a rendezvous in Queen Charlotte's Sound, New Zealand. Furneaux had copies of Cook's charts to find it. He was a good seaman. There was no need to worry over him.

The *Adventure* had orders to make for the east coast of Tasmania first and did so. But she remained there a few days only and did not find out that the place was separated from New Holland – a minor point at the time. Nobody was planning anything in Tasmania, except perhaps the French. To Cook nothing in discovery was a minor point: but when his turn came to sail north, heavy gales and the great seas that ran with them forced him to bear away. He must take these awful seas on the beam to reach Tasmania, rolling like huge watery hills and breaking dangerously as they flung their gale-tortured crests towards the sky. The *Resolution* could ride them out better if they came at her from slightly behind her: she was already three months at sea from the Cape, and no scurvy-grass had offered from Antarctic ice. So Cook, too, bore away, and made for the south island of New Zealand. One hundred and seventeen days out from the Cape of Good Hope, having sailed 11,000 miles – much of it where no other man had ever been and all of it among the most dangerous seas in the world – Cook brought the *Resolution* to anchor in Dusky Sound, a sheltered bay in New Zealand's south-west corner. The only casualties to scurvy were the few surviving sheep: not a man was seriously sick, or had been. The sheep, poor beasts, would not eat 'sour krout' or drink 'rob' of lemon: their scurvy-loosened teeth were rolling sideways in their mouths and at first they could not even eat the luscious greens that Dusky Bay provided. Poor sheep! When they did recover, some of them promptly ate from a poisonous bush and died.

Dusky Sound was such a place as would have delighted Banks: the *Endeavour* had had to sail past the entrance. Now Lieutenant Pickersgill found the ship a lovely inner cove which bears his name – a tiny, sheltered place, quietly beautiful, where the yards reached in among the branches of tall trees and a fallen trunk made a convenient jetty. The idyllic spot, with its charming and utterly unspoiled air as if a benevolent Nature had designed it for the succour and rest of seamen up from the Antarctic and 'shrieking fifties' storms, offered fish, seals, and some game, firewood, fresh water, good timber for repair of spars, decks, and hull – everything but fruit and vegetables. It was perhaps too far south for fruit: nobody had ever planted vegetables there. Cook made up for this by sowing seeds brought for just that purpose. But he could not wait for produce from them: this was not the rendezvous. He was delighted with the Sound where even the ship's cats prospered, though they fell hungrily upon far too many unsuspecting birds. The chafed back-stays and lifts and the foreparts of all the tops were protected by the skins of fat seals, the carcasses whereof had first provided good steaks, and their fat was rendered down for lamp-oil. There were plenty of whales offshore, as there had been also in the more open Antarctic waters. They let these be. The Greenland men recognized most as heavy rorquals of the kind that sink when dead and so were useless to eighteenth-century whalers. The *Resolution* had no whaling gear anyway, but this news of whales was to create great interest later in Dundee, Aberdeen, London, Whitby, New Bedford, and other whaling ports in Europe and New England.

The *Resolution* passed six weeks pleasantly and usefully in Dusky Sound. By mid May she sailed into that other fine harbour at the other end of the island, Queen Charlotte's Sound. At first they could not see the *Adventure*, for she was rigged down. Furneaux did not yet appreciate the nature of his Captain. He had snugged the smaller ship down to her lower-masts and moored her for a stay of months, under the pleasing impression that he was now in winter quarters. After all, it was the winter season: Antarctic voyaging had been tough: he was used to senior officers who did not drive themselves.

When Cook did see the *Adventure* lying like a rigged-down hulk he was shocked: when Furneaux reported aboard he was horrified, for there was scurvy aboard too – and that after six weeks in harbour where 'sellery', scurvy-grass, and other such useful items were to be found in abundance.

It was no part of Cook's plan to fritter away the morale of good crews by months of idleness in a rigged-down hulk nor to have his men die avoidably from scurvy. He stopped the extra rum rations his junior was dispensing (to help while away the time, perhaps) commanded the re-rigging to be begun forthwith, showed the *Adventure*'s boats where to fill with good 'sellery' and ordered the daily consumption of these fresh things boiled with wheat or oatmeal, or portable soups, for breakfast, and likewise boiled daily with pease for the midday meal no matter what else might also be eaten. Boats were sent out daily to catch the excellent fish: some of the *Resolution*'s seamen were sent across to help the scurvy-struck crew of the *Adventure* rig their ship again as quickly as possible: stocks of anti-scorbutic grasses were laid in for both crews and the few remants of their livestock.

The *Adventure*'s men, now properly looked after, soon recovered, but it took over a month to overhaul the rigging and set it all up again properly. In the meantime the *Resolution* remained the taut ship she always was, friendly relations established with the Maoris by the benevolent Furneaux (whose only fault was a great capacity for the easy life which, after all, was a peace-time characteristic long accepted in the Services of those days) were fostered and continued. Furneaux had set up some fine gardens in pleasant spots ashore, with promising crops of potatoes, turnips, carrots, parsnips and so forth grown from seed. There was not time to wait until these could be harvested, so they were made over to the Maoris, who seemed delighted. They were even more delighted with a gift of breeding hogs and goats, provided by pooling the livestock of both ships.

At last, early in June, the *Adventure* was once again scurvy-free and restored to her rigged state. The two ships unmoored, got all the anchors up and catted or stowed, and passed with a favouring wind out to sea. They were bound to the eastwards to look once more for sign of Dalrymple's *Terra Australis*, first in that part of the South Pacific where the *Endeavour* had not been – between New Zealand and mid-way to the Chilean coast – and afterwards to make a wide circular sweep towards the tropics to sail in unsearched waters there, not in quest of any continent where Cook knew there was none, but of islands large and small. Thence, in good time for the next southern summer – that of 1773–4 – back to New Zealand for a brief refitting spell before making once more for the Antarctic.

It was, as usual, a bold plan for a bold voyage which could also be a trying one – at least in far southern latitudes.

Cook explains himself in his Journal, where he seems at times to justify himself against future or perhaps present criticism, for this idea of mid-winter Roaring Forties–Shrieking Fifties voyaging was new. 'It may be thought by some an extraordinary step to proceed on discoveries as far South as 46 deg. in the very depth of winter, for it must be own'd that this is a Season by no means favourable for discoveries. It nevertheless appear'd to me necessary that something must be done in it,' or he might find himself with not enough time to finish his enormous job. And, if he did come upon a real piece of land in these unknown or little-known seas, he would then have part of the summer to explore it, and time enough to sail along the Pacific sector of the Antarctic as well. He had, he concludes, 'little to fear, having two good Ships well provided, and healthy crews'. Besides, if he discovered nothing, at least he would 'point out to posterity that these Seas are to be navigated and that it is practical to go on discoveries, even in the depth of winter'.

All typical Cook, of course. Off he went, beating against easterlies in Cook Straits, the *Adventure* going a little better to wind'ard. Clear of the land, soon the wild westerly gales found them again, howling in the rigging and bringing up a vicious wintry sea in which the ships 'labour'd much', and the force of the rollers smashing at the rudder jerked the tiller of the *Resolution* violently, flinging the helmsman right over the wheel – a nasty, bone-crushing habit sailing-ships running their easting down were to maintain until the end of the era. By good fortune, the helmsman so rudely thrown 'resembled a seal in substance and make' and so thwacked the deck like a 'bag of blubber'. The officer of the watch had immediately jumped to the wheel and held the spokes (or the rudder, banging wildly, might have smashed itself, or at least broken the tiller to pieces): the 'bag of blubber' picked himself up, no bones broken, and going back to the weather side, carried on grimly steering. Sudden mulish kicks from a square-rigger's wheel were always a risk, especially when running before a heavy sea. To guard against them, relieving tackles were rigged to restrain the tiller's tendency to jerk when a sea slapped the rudder, and the weather helmsman – he at the windward side of the wheel – wore a safety harness rigged stoutly to the deck, to hold him down. Men were sometimes thrown right overboard, or broke their limbs when hurled over the spinning spokes and smashed down on the wheel grating or the deck. It was usual to have at least two men at the wheels of large sailing-ships, one each side. Steering-wheels were controlled from *beside* not behind them, to give a man a stronger grip. Cook often had two men to the wheel, too. After this incident, the officer of the watch sent an extra hand to the lee side of the wheel.

In July, still south of Forty, Cook was setting studding-sails, and putting over a boat to row Astronomer Wales over to the *Adventure* following him, to compare chronometers one of which had stopped. A day or two later one of the goats foolishly fell overboard: a boat went back and saved it, but the poor thing died soon afterwards, perhaps disconsolate with the ship's endless motion, the cheerless decks, and a bellyful of cold sea water. Mr Forster Senior, scowling down below where he was flung about in his 'mouldy and damp' cabin, was full of complaints as usual, though even he had by now some reluctant admiration for his Captain. The stay at New Zealand had been a surprise to him: his son loved it.

This wintry sailing was all right for Cook and his lads, perhaps, and they made no complaint. But it was tough on goats, chronometers, and passengers.

CHAPTER ELEVEN

Among the Ice and the Islands

THE circular passage that Cook was making as part of the overall plan of this second voyage was the ideal method for a square-rigged ship to investigate the South Pacific ocean. As planned, it meant sailing eastwards with the west winds, north with the variables, west again with the tropic's trades among the atolls and the islands, and then completing the circle southwards at will, first with the variables and then taking the west winds on the starboard beam. In due course, to reach the South Atlantic, it was necessary only to run with those same west winds, and swing on past the Horn.

Of course, these various winds did not turn themselves on as a ship reached the zones where they could be expected, for life is never as simple as that. But they could be relied upon to perform sufficiently in due course, for those skilful in seeking them and patient in waiting for them. A ship might make any number of these broadly circular sweeps and so cover the South Pacific thoroughly. She must avoid the odd hurricane season here and there – the cyclones of the Coral Sea and much of the Melanesian Pacific in January to March, and the occasional hurricanes among the Tuamotus. These can bring up a sea that washes right over low atolls and islands in their path, and are no help to sailing-ships at all. To heave-to in the path of one, though sometimes the only course possible, is a dangerous thing, for then the ship may be washed on to a reef or atoll, too. There are the rusted hulks of wrecked sailing-ships on reefs in the South Pacific to this day. The correct procedure is to heave-to in such a way that the ship may be worked out of the storm – they are corkscrewing, circular things – instead of being sucked further into it.

I sailed my ship *Joseph Conrad* along Cook's way, from Cook Straits towards Tahiti, in the winter of 1936. She was a near-perfect full-rigged ship, as safe as such a ship could be and in good state and trim: but I was pleased to get out of the west winds zone where the ship stumbled and pitched and rolled as the gales screamed in the rigging and the sea flung her along. Even the royal albatross, wandering up from Campbell Island, gazed on her with mild astonishment as the three mastheads, circling madly in the sleet-filled sky, seemed to reach out for them, and the violent eddies in the lee of my close-reefed tops'ls set up such an additional tumult in the wind as we rushed by that it caused even those most expert gliders to lose lift and control, and sometimes come crashing in the sea. Once a huge sea dropped it murderous crest over my weather quarter, washing away the seaboat and taking a small poop ladder overboard: some fool ashore later reported us 'missing'. We were received at length at Papeete as if we were back from the dead, though my whole passage (from the Lousiades) was seventy-seven days.

Cook ran the *Resolution* and the *Adventure* well to the east of the longitude of Tahiti, the better to examine these southern latitudes. On a quiet day of long, greasy swell, he put out a boat and visited the *Adventure*. (Boat-work and visits between ships in company was standard practice then: there was often time.) The ships were less than a month at sea from New Zealand but the *Adventure* already had twenty cases of scurvy. The *Resolution* had none. Furneaux had been doing his best, but the rigorous adoption of Cook's dietary methods was not enough. *All* his methods had to go with it – cleanliness at all times, airing the quarters and the hold with iron pots of charcoal fires, insistence on the consumption of the full issue of anti-scorbutics based on good example set in the great cabin, unrelenting attention to detail in the care of quarters and men whether the men liked it or not. This sort of thing, brought from the *Endeavour* and well understood and fostered, had long been normal drill in the larger ship. Furneaux was no James Cook: certainly he had no help from 'Murduck' Mahony, his slothful cook.

'Departed this life Murduck (sometimes Murdoch; on the muster roll as Mortimer) Mahony Ship's Cook,' is the brief record in the *Adventure*'s log, but the astronomer Bayly, in his journal, is frank enough. Murduck alias Mortimer Mahony, he says, 'died of the Scurvy, he being so very indolent and dirtily inclined there was no possibility of making him keep himself clean, or even come on deck to breathe the fresh air'. Poor 'Murduck' was a born eighteenth- (or nineteenth-)century sea cook, obviously. Some reluctance to exchange the warm fug of the galley for the wet and windy fore-deck may be understandable, as in addition to its other disadvantages that end of the ship must have stunk abominably of goat and other livestock, whenever the wind was favourable.

Cook soon found a replacement for the departed Murduck, having an excellent seaman ready and willing who had, apparently, recently lost several fingers but was otherwise whole, though aged, and was well indoctrinated in the *Endeavour* standards. He was rowed across and set forthwith to work: but the scurvy cases in the *Adventure* continued to increase. By the end of July, with the ships lolloping about in the Horse Latitudes making twenty or thirty miles a day (and that sometimes sideways) she had twenty-two men sick, and even the *Resolution* had a few mild cases. These were promptly dosed with brewed wort, and all the rest of it, not forgetting Baron Storsch's 'marmalad of yaller carrots' and – perhaps to avoid this sort of thing – as promptly reported back for duty. Cook altered course to make towards Tahiti, that sure haven of refreshment, but the trade winds refused to blow and the scurvy in the *Adventure* would not die down. Furneaux himself was reported on the sick list in one of the morning hailings, but his sickness was diagnosed as gout – poor fellow, for that must be most unpleasant on a rolling quarter-deck. In the quiet weather there was more inter-ship visiting. Mr Forster Senior and the artist Hodges pulled with the officers across to the *Adventure* and once stayed all day. Such boat-work was no new thing. Whenever there was a chance they were at it in the Antarctic, too.

At last the trade winds found them and they sailed along the old westwards course before the wind, near the Tuamotu islands. Furneaux broke out some Devon cider for the worst of his scurvy cases – a third of the crew were affected when the trade came – and it did them good.

By mid August the mountains of Tahiti were in sight. Then it fell calm, for the trade wind was upset by the high land.

That calm all but wrecked not just the *Resolution* but both ships, for it was the Queensland Great Barrier Reef situation all over again. Helpless before the set and the scend of the sea, unable to sail without wind, the ships drifted towards Tahiti's fringing reef. All the boats were out trying to tow, but this was hopeless. The *Adventure* was short of fit men and so Cook had to lend some from the *Resolution*, leaving both ships short-handed. There had been a little sailing breeze in the night, with perfect sailing conditions. Towards midnight, having given his instructions to keep the ships on a course well clear of the reef, Cook undressed and turned in as was his habit, for he had perfect confidence in his officers. It was clear weather. There was no need to get close to the reef. But on this occasion there was undoubtedly an error somewhere: the standard of watch-keeping somehow slipped. In the beautiful night the lovely mountains of Tahiti so close by made the perfect back-drop for thoughts, perhaps, of romance and soft adventure, of which the rich promise after long hardship now lay so close at hand. Yarning with quiet voices by the break of the quarter-deck away from earshot of the captain's sleeping cabin – and away from the wheel too, out of sight of the binnacle – the lieutenant of the watch, the Master's Mate, the quartermaster might have been lulled from their usual strict attention to the details of duty. Some misunderstood, mumbled compass course at the hourly change of helmsman, a mis-heard repetition to the watch-keeper, a failure to appreciate the slow and gentle dropping of the very light wind until the ship stood silent there, all sails and reef-points quiet, while the stealthy current, all unperceived, edged her slowly, steadily towards the reef, and the height of the tide kept the warning surf there for the moment quiet. Then the dawn, so gently beautiful, so peaceful as if all evil were gone from the world forever: and a sudden tumult soon of fleets of canoes full of the happy brown-skinned Tahitians striking the sea with swift paddles, shouting '*Taio! Taio!*' as they came with their gifts, their welcome, their fruits and coconuts to trade. Laughing, smiling, shouting, scores of beautiful young women, bare to the slim waist, clambering up the sides, clinging to the chain-plates, swarming over the channels, leaping bare-limbed over the rail. '*Taio! Taio!*'

'Lee fore brace! Down helm!'

The sudden, impervious voice from the companion was Cook's, taking in the situation at a glance. 'Bring her up to the wind!'

But there was no wind.

'Clear the hawse! Tumble up the hands! Away the boats! A sounding, Mr Gilbert, if you please!'

But the Master's men could find no bottom. The ship could make no way. The boats could not hold her. And no wind came.

Slowly the *Resolution* edged in towards the island and its reef, changed upon the instant from long-sought haven to dangerous menace intent on murder. The *Adventure*, slightly further off-shore, drifted on a converging course. A light bower anchor was outboard with thirty fathoms of cable, to bring the ship up in the hoped-for soundings if anchorable bottom preceded the steep-to reef. No bottom. No usable wind. No manageable way. Not a cable ahead with the boats, though the sailors sweated. Now the surf began to smash upon the coral as if the reef was licking its hard lips in anticipation of the unusual pleasure of a ship to grind and smash to pieces.

Cook noted that, as the movement of the water set the ships towards the reef, very slowly there was a slight change of direction to the end of the reef, a point which if they cleared would lead to open water again. Which would come first, the reef's end or the ship's grounding? It looked like touch and go! Still the straining boats could give the ship no headway.

The morning dragged on, hot, windless, stifling. A slight break showed in the reef. Get in there? The boats could tow in with the set, perhaps: never out against it. But the Tahitians said there was insufficient water. It was a break in the coral, not a channel through. A quick check from a boat pulled rapidly in proved this. Instead of offering brief and perilous chance of safety, the opening caused instead an indraught right into the reef. On lurched the ship helplessly, silent, graceful, every detail from proud cut-water and jibboom-end to carved transom and open stern windows mirrored perfectly in the still water; anchor finding no bottom to grip though now down at 100 fathoms, boats useless, sails lifeless like listless white sheets hanging dead and silent in the sun. No bottom! No bottom!

Then she found the bottom with her keel.

Her forefoot smashed down with a jar that shook the rigging. The bow rose, the stern pounded.

'Goddam!' said Captain Cook.

No ship could stand much of this.

The sea which had been so quiet now broke upon the reef with a violence heightened by the fears of the listeners: but the *Resolution*, held for the moment by the light bower anchor, avoided broadside grounding. She was making no water, not leaking – yet. The bottom just there must be sand. The swell upon the reef lifted her with its rising as well as pounded her with its hollows. The decks filled with sweating, brawny men as the boats were recalled alongside. The *Adventure*, for the moment, was still safe. An anchor seemed to be holding her.

'Carry out a kedge!'

Impeded by the crowd of merry Tahitians cluttering the decks, the sailors got light anchors over and carried them out with swift efficiency on long hempen lines, the tall figure of Cook in the bows directing them. Sails were run down, gathered loosely in their gear, left. Back came the kedge-carrying boats, the men pulling for their lives.

Still no water in the well: and thank God for that. They could see the bottom now. It *was* sand.

'Man the windlass!'

Swiftly the handspikes flashed, in came wet hemp. The lines tautened. The windlass strained. The ship moved ahead! At that very moment the light air came again. The sails flapped, slightly at first, then with the imperious sound of fitness for their work.

And still the *Resolution* was not leaking.

Sheet home! Man halliards! Ease lee braces! A hand aloft to each mast, overhaul gear!

Great hams of hands grasped halliards, great horny feet

stamped the decks: up sang the yards: down flapped the sails. The ratlines and all the shrouds shook with the swift passage of climbing boys.

At this same moment the *Adventure* got under way, too, thanking God for the breeze. But while she stood briefly unmanageable until she could get her anchors home – or try to – there was imminent risk of collision in the confined area just off the reef. The two ships passed within feet, the *Resolution*'s men standing for'ard with stuns'l booms to fend off, the *Adventure* axing away three anchors to get out of the path. Before they came clear, the two ships were so close that 'a tolerable plank would have crossed from gunnel to gunnel'. After that sudden, scaring brush with shipwreck, Cook went round to the old *Endeavour* anchorage at Matavai Bay, and the *Adventure* followed. Here they moored with their surviving ground tackle, secured all sail, and got on with trading for fresh food.

It had been a bad morning – nothing like as perilous as the Queensland Barrier Reef which had caused almost mortal damage, but bad enough: worse, perhaps, because it should have been avoided. But any ship-master may have an accident and the South Sea islands were notoriously dangerous throughout the days of sail. It is the actions of a Master after his ship gets into trouble that show the stuff he is made of.

We have an independent witness of Cook's worth on this occasion (if one were needed). The Forsters had induced their captain to allow them to invite another scientist aboard at the Cape, a young Swede named Anders Sparrman, one-time pupil of Linnaeus and an excellent botanist. Sparrman afterwards wrote a useful book in Swedish, which had to wait more than a century and a half for publication in English.*

'None but a seaman can realise how terrible was the sound of the waves breaking on the coral reef so near to us, mingled with the shouting of orders and the noise of the operations our dangerous situation made necessary,' he writes. 'But even in my anxiety, I drew no small satisfaction from observing the rapidity

* *A Voyage Round the World with Capt. Cook in H.M.S. Resolution*, by Anders Sparrman (London, Robert Hale Ltd, 1953, after Golden Cockerel Press Limited edition, 1944).

and lack of confusion with which each command was executed to save the ship. No one seemed aware that he had been working for hours under a burning sun, the thermometer at 90° in the shade ... I should have preferred, however,' he continued with odd priggishness, 'to hear fewer "Goddams" from the officers and particularly the Captain who, while the danger lasted, stamped about the deck and grew hoarse with shouting.'

An odd Swede, Mr Sparrman: did he think a few ladylike gestures would help to get the ship out of peril? This was man's work: he gave out a few 'voice trumpets' from the cabins to ease the hoarseness, and wisely kept out of the way.

He does throw some light on the effect of the strain on Cook, who indicates nothing of it in his own journal, or of the state of his personal health. 'As soon as the ship was once more afloat, I went down to the Great Cabin with Captain Cook who, although he had from beginning to end of the incident appeared perfectly alert and able, was suffering so greatly from his stomach that he was in a great sweat and could scarcely stand,' writes Sparrman, who adds that he persuaded the Captain to take a glass of brandy which did great good. Where the rum, brandy, arrack, wines, and so forth flowed so freely, Cook rarely touched any of it.

As for Mr Sparrman, his shipmates had a smile when a couple of 'guides' at Huahine seized him later, stripped him, and made off with his clothes. Cook told him severely that he had been most imprudent, going off alone with a couple of Society Islanders, and warned him not to do such a thing again. Perhaps he smiled inwardly, too; though there is little in the records that whole voyage to indicate that he found any of the three scientists pleasing fellows. The Forsters were for ever wanting foreknowledge of his plans in detail (he would have liked some such knowledge himself at times, especially in the Antarctic) and Forster Senior came with constant complaints about his cabin, the crew, the livestock, and anything else he could think of. As for 'Spearman' – as Cook calls him – it was obvious to him soon after the Forsters induced him (and paid him a good salary out of that £4,000) to join at the Cape that the Antarctic held little promise for botanists: after that, he confined his exertions to New Zealand and the islands.

The ships lay moored in Matavai Bay some weeks while the crews rested and the *Adventure*'s too many scurvy patients slowly recovered. The Tahitians were as skilful and persistent pilferers as ever. There had been some political troubles, apparently, and provisions were scarce. So Cook soon sailed on westwards to the Society Islands – to more great welcomes, more pilfering and sometimes serious thievery and, more usefully,

Resolution *and* Adventure

excellent supplies of fine hogs, fat fowls, yams, coconuts, plantains, 'bananoes' and the rest. The ingenuity of the thieves descended even to cheating on the coconuts, by first drinking the juice, then filling them with water and skilfully sealing them again. These activities were carried on with such a playful air of childlike innocence, and restitution made so cheerfully on discovery, that it was quite impossible to be angry for long and utterly hopeless to get any of them to mend their ways. Their thievery was continuous, determined, and ingenious: it seemed also to be an amusement. Even tricing the occasional habitual purloiner up to the shrouds and giving him a dozen of the best made no difference, for they accepted this knowing it was just, and went away to return another day, taking the cat-o'-nine-tails with them if possible, or anything else handy.

Ship's punishments of this sort were reserved for natives caught stealing irreplaceable ship's gear, like the capstan or a spare anchor, one of the sand-glasses, important books, or the furniture in the great cabin (which an alleged minor chief once tried to pass out through the stern windows to confederates in a canoe below). Any member of the crew who tried to cheat the Polynesians was punished in the same manner. Nothing made the slightest difference, not even a ceremonial chastisement which the exasperated Cook once laid on ashore in the Society Islands, with the marines drawn up in full regalia and speeches about equal justice for all and the evils of thievery etc., before the burly bo's'n swung his cat. The natives were not ordinarily thievish: but there was no other source for the wonderful things the ships brought to them. About the only things no one tried to steal were the masts and the bagpipes.

Cook was always humane, always aware that he was the interloper, that these Pacific peoples would have been better off left alone. Above all else, he strove for good relations with them, for amity and mutual respect: but he had ships and people to look after, an arduous voyage to make, and the half of it was not yet over. If only there were not all this thieving! At Tahiti many wanted to sail with the ships, and chiefs wept at their departure. At Huahine the old chief, in tears of joy, insisted on setting up elaborate ceremonies to receive Cook with fitting respect, and produced hogs by the hundred and roots and fruit by the ton.

At islands to the westward which Tasman once had briefly visited, there was such harmony and pleasant trafficking with the Polynesians there that Cook named these the Friendly Islands and the name (though not always thought, perhaps, wholly appropriate by other pioneers) and disposition have stuck. A local chief known to Cook as Otago so filled the ships on sailing with fruit, yams, hogs, and poultry that 'we had to keep plying under tops'ls until the decks were cleared'. Slippery little hogs scampered about the decks as swiftly as Mr Banks's long-gone greyhounds and poked their snouts through the open taffrail, looking at the islands. Poultry of gorgeous plumage (but as the sailors soon found when they tried to organize cock-fighting for a dogwatch diversion, poor fighting qualities: perhaps they had absorbed the general friendliness too) ran with flapping wings out of the seamen's way as they put the ship about. This was a manoeuvre which demanded swift access to almost every piece of the ship's gear. Plantains, bananas, bunches of yams, tied-together coconuts cluttered the coiled halliards and braces, hung from every cleat and belaying-pin, filled the stowed boats, hung from the quarter-deck rail. It was a cheerful confusion, brought quickly under control.

Otago himself stood aft in the midst of it, with Cook, tearfully requesting the captain's return with, if possible, the gift of a uniform. At length the decks were clear enough for the ship's gear to be properly workable: Otago and suite left by canoe: Cook altered course away from the wind, set all sail now the mariners could get at the halliards, and stood away for New Zealand. It was already the first week of October 1773: it was time to head once more for the Antarctic.

First, it was essential to spend a day or two in that perfect base, Ship Cove, in Queen Charlotte Sound. Cook had the *Resolution* safely moored there a few weeks later. Once again the *Adventure* was missing. A heavy gale on the New Zealand coast parted the ships: but Furneaux had only to lie it out – the wind was offshore – and then sail back again. Weeks passed and he did not come. Cook stocked the *Resolution* with celery, scurvy-grass, and some of the vegetables he had planted (for the Maoris, still eating one another after bloody combat, had paid little attention to these): took aboard firewood and good timber, renewed stocks of fresh water, caught fish by the boatload, and distributed some of the Friendly Islands' hogs, cocks and hens in the hope that the Maoris would let them prosper – a rather vain hope, for the Maoris though 'manly and mild' to Cook, were apparently in (or ready for) a state of constant imminent warfare. The land they might have tilled, the hogs they might have tended could be another's on the morrow, with themselves broiled as a side dish: therefore a hog in sight was a useful meal, for life was short.

On 25 November Cook unmoored and sailed, leaving instructions in a buried bottle for his absent junior who might have been blown away towards Cape Horn, partially dismasted. There was no more time to waste: a sailing-ship might stay in Antarctic waters only until March, and there was the whole great sector from south of New Zealand towards Cape Horn still to be investigated.

The *Adventure* did finally beat into base, bedraggled and sea-weary, a few days after Cook left, but it was too late then to think of rejoining. Cook was on his own. After a rest and refit – and the loss of a boat's crew, a couple of midshipmen and eight men, killed and eaten by the Maoris – Furneaux sailed away towards Cape Horn, the Cape of Good Hope, and England. He crossed the Pacific by a far southern route no one had then investigated, adding another great area to the land-free sea, looked again in the far South Atlantic for the elusive Cape Circumcision, and was back in England twelve months before the *Resolution* – just in time to pick up a command he probably did not want, the new frigate *Syren*, sixth-rate of twenty-eight guns, fitting out for the Wars of the American Revolution.

In the meantime Cook stood away once more across the zone of the Roaring Forties into the Shrieking Fifties and towards the Antarctic waters round and south of 60° South. Again the 'ice islands', the growlers, the floes and at length the pack-ice added to the sailing hazards and the crew's hardships. Again the sails and all the running rigging froze, and the Fearnought clothing of the men – well-worn by this stage – stood like frozen armour on them and their knees broke the ice with an odd crackling as they walked. Fogs, gales, blizzards added to the dangers. But morale remained high. Christmas Day – second that voyage – was spent in the area of 67° S., across the Antarctic Circle, the 'wind northerly a strong gale with a thick fog sleet and snow which froze to the Rigging as it fell and decorated the whole with icicles ...' – a wonderful Christmas-card scene, but grim for the seamen.

They celebrated in the usual manner by first snugging down the ship so she could come to no harm and then getting happily and uproariously drunk, and staying that way as long as possible. Sparrman, counting 168 icebergs large and small in sight from the ship, reports that the officers who the day before were declaring themselves disconsolate with the Antarctic monotony and ready to change the frozen scene 'for the fiercest sea battle, broadside to broadside' now loudly publicized their willingness to remain in the Antarctic even if wrecked on an iceberg (a real enough prospect) where they would 'die happy with a rescued keg of brandy in their arms'. Inside their skins was more likely their real meaning: but Sparrman was impressed. He was sure that nobody aboard, even when sober, would vote to discontinue the voyage no matter what the risks or hardships: rather they 'would have made every effort to press on towards the South, even if it had been a question of hoisting the British flag at the South Pole itself'. Mr Sparrman was obviously absorbing the spirit of those days, when a temporary 'drunk' was the only relief the seaman had.

As for trying to reach the South Pole, it was only the absent Banks who had ever mentioned it. He had some odd idea of standing on it and swinging himself playfully around 360 degrees of longitude with the earth's rotation. The pole lay impregnable at the heart of the ghastly Antarctic mainland – that frozen *Terra Australis* so long and defiantly *incognita* – and no man was to stand upon it before the Norwegian Roald Amundsen raced there with his dog teams 135 years later. As for Cook, after reaching the latitude of 71° 10′ South and sailing inside the Antarctic Circle for days, an impenetrable icefield blocked his path both south and east. So he turned away. This was on Longitude 107° W. (within a mile or two), not far from that part of the Antarctic known today as the Walgreen coast of Marie Byrd Land, well west of Graham Land. In the pack Cook counted 'ninetyseven ice hills, besides those on the outside, many of them were very large. They looked like a ridge of mountains rising one above the other till they were lost in the clouds.' And here, against this background of chill grandeur, at last Cook takes a slow look at himself, and confides to that prosaic journal of his a personal thought or two which throw some light on the man:

'I who had Ambition not only to go Farther than any one

had done before, but as far as it was possible for man to go, was not sorry at meeting with this interruption as it in some measure relieved us, at least shortened the dangers and hardships inseparable with the navigation of the Southern Polar Regions: Sence therefore we could not proceed one Inch farther to the South, no other reason need be assigned for my Tacking and Standing back to the North . . .'

'About Ship! Stand by to go about!' was the immediate order. Young Midshipman George Vancouver, at the moment off-watch, raced to the jibboom-end and stayed there in the bitter cold, clinging bare-handed to the fore topgallant stay, determined that he, personally, should reach the farthest south, if only by the length of bowsprit and boom. The ship was sailing in brash ice and small floes. The wind was easterly, the ship on the port tack, headed south.

'Lee-oh! Hard down the helm!'

The eased fore-sheet and the jib-sheets set the head sails to flapping violently as the ship shoved her valiant nose to wind'ard.

'Ne plus ultra!' bellowed Master Vancouver, waving his balaclava at the wind, and climbed slowly back aboard to boast for ever afterwards that he had been closer to the South Pole than any other man on earth. And so he had, by possibly twenty or thirty yards.

After this, Cook made a steady careful course towards the north, away from Antarctica: nor did he cross the Antarctic Circle again. By sailing the way he did to and from New Zealand early and late in 1773, he missed the only place where the frozen continent was sometimes vulnerable – the Ross Sea. He had turned north-eastwards off Adelie Land, not far from the Balleny Islands, and come south again somewhere near 160° W., well outside and to the east of the entrance to the Ross Sea. The heavy westerly gales, indeed, forced him to go that way, as the *Resolution* sailed slowly to the south'ard from New Zealand. In December – the time Cook was there – the field of pack-ice and frozen-in growlers and icebergs usually (but not always) disperses from the outer waters of the Ross Sea. When that happens, a ship may sometimes sail right up to and along the face of the Great Ice Barrier at the edge of the Ross Ice Shelf, beyond 78° South – the farthest south that a ship may get, and almost 500 miles farther south than the *Resolution* managed.

It was another sixty-six years before this was discovered by the ice navigator, Captain James Clark Ross, R.N. (later Sir James Clark Ross), with two wooden square-riggers – the *Erebus* (370 tons) and *Terror* (340 tons) – not unlike the *Resolution*, but ice-strengthened. Early in January 1841, Ross, already an experienced Arctic ice pilot, came to pack-ice on 174° E. inside the Antarctic Circle. He made boldly into the pack, which was loose and much softened by a good summer, and to his surprise worked through to the South in five days. To his far greater surprise, beyond was open sea, bounded to the west by a vast, mountainous land. This he called Victoria Land, for Queen Victoria, and sailed south until brought to a stop by the perpendicular wall of the Great Ice Barrier which surely must be one of the most astonishing sights in the world.

Here in the Ross Sea the Antarctic mainland was vulnerable: here ships could get within 700 miles of the Pole. At the Barrier's western end, the Victoria Land coast formed the useful harbour of McMurdo Sound. From this sound, and a sort of crack in the Great Barrier called the Bay of Whales, the great overland sallies were made towards the Pole – by Shackleton, Amundsen, Scott – and the airman Richard Byrd based here too. Ross came from Hobart, in Tasmania, to wind'ard of his sea and therefore with a slightly easier run than Cook's from New Zealand.

It is unfortunate that this dip south of the Antarctic and sole hope of accessibility by sea coincided with one of Cook's sweeps to the north, and that he then had to make south across the west winds again from a base which put him to leeward. He could not discover everything, of course: but one regrets his lack of luck, perhaps, in not coming on the Ross Sea. The sight of mountainous Victoria Land and that Barrier face would have been momentous occasions of which he was worthy. One would have liked the Forsters and Anders Sparrman to see these things, too, for they did indeed suffer much cold boredom and saw no Antarctic land at all. They could have landed at McMurdo

Sound, and they would have had the delight of 'discovering' the Emperor Penguin, that dignified, black, white and gold four-footer who lives in the Antarctic the year round.

In the meantime, having now proven that there was no unknown southern continent except whatever might be found locked beyond the ice, Cook continued to fight the west winds to begin what was to prove his most extensive sweep of the South Pacific and the most productive voyage of discovery there ever made. Taking as his guide the account of Pacific voyages assembled by the energetic Dalrymple, he made north first to look for the alleged 'mainland' sighted by Juan Fernandez towards the end of the sixteenth century and accepted by Dalrymple (if by no one else) as the eastern extremity of *Terra Australis*. It was immediately obvious to Cook that, whatever this place might be – it is not to be confused with the real Juan Fernandez Island, then well known – it had no western area, for he sailed over it. There was nothing there at all – no land anywhere.

In the meantime, slogging away determined to sweep these wills-o'-the-wisps from the sea, driving his ship and himself without let-up, Cook nearly died. He suffered from a 'bilious colic', he writes in his journal: every other journal aboard – many survive – has him on the point of death. Poor man, his stomach had been in knots for weeks. Racked by violent fits of hiccoughing which nothing seemed to ease, throwing up whatever he tried to eat, so weak he could not stand and unable at last to accept any nourishment at all, there was only one possible end even for his tough constitution. His whole digestive system had broken down. Tormented for years by the sea diet of salt horse dosed in nitrate, greasy and repulsive salt pork, all manner of frightful 'anti-scorbuticks' consumed with gusto for the sake of example to the sceptical and conservative mariners – all this relieved at odd intervals by temporary excess of tropic diet, poisonous fish, razor-backed hogflesh possibly diseased, unripe and over-ripe fruit, the astonishing thing was that anyone's stomach could stand such a constant period of strain and overwork at all. Even with the sea diet of today stomach ulcers remain a deepsea merchant seaman's occupational hazard, though stress is an enemy here too. Cook knew stress in abundance. He probably had ulcers: he certainly had a malfunctioning stomach and no hope of relief. He became painfully thin, and his racking hiccough was almost unbearable. The surgeon did all he could: but what use was that?

In this predicament, with the ship over two months at sea and no animals left or poultry or fresh food at all, and the patient long weary of petrel soup and boiled penguin – ill-tasting concoctions both – the elder Forster stepped into the breach. He had a pet dog, a Tahitian animal, plump, and cereal fed. Could the captain, perhaps, face a dish of soup made from this if it were sacrificed? The captain could, feeling that at any rate nothing could make him worse. The soup was made, the captain ate it and kept it down, ate a bite or two of boiled leg of dog, kept that down and, from that moment, began to recover. Word rushed through the ship: smiles broke upon the anxious faces of every man and boy aboard – smiles for Cook, smiles for irascible old Mr Forster too.

After 103 days at sea, with Cook still so weak he could scarcely hold a telescope steady, they came to that most odd Pacific island, the one called Easter. The natives were friendly and the massive, mysterious stone heads most interesting, but there was little sustenance: Cook soon pushed on. There was no sign anywhere of large lands. He made towards the Marquesas, not seen since the Spaniards from Peru passed by in 1595 and ill-charted them. Cook found them, traded with the fine-looking, pleasant Polynesians there, admired the islands, and passed on to that sure haven of refreshment, Tahiti. The Marquesan pigs had multiplied too much, apparently: though abundant, it took fifty of them to give all hands a meal. Tahiti could do better than that, and did. Tahiti and the Tahitians were amazingly prosperous and as friendly as ever – and the people as light-fingered.

Here Cook saw the stirring sight of a fleet of several hundred war canoes – big, double canoes, sixty and eighty feet long, manned by 8,000 braves led by imperious aristocrats in colourful garments of war. Well, perhaps not quite war: the fleet went skilfully through intricate manoeuvres of amphibious warfare, dashing in turns at great speed upon the shelving coral strand with a great play of banners and warlike shouts. What a sight this must have been! The campaign, according to the chiefs, was against Tahiti's neighbour Moorea: but the 8,000 braves in the 400 war-canoes never sailed – at least, not while Cook was with them. There were hints that they might have liked support from the *Resolution*'s guns, and this could not be forthcoming. King George's warships were not for fostering brief inter-island wars.

Tahiti, Huahine, Raiatea, Tonga (the Friendly Isles, as friendly as ever) – this was now a familiar round, a smiling summer South Seas cruise which, under their humane, for ever just and considerate leader, Cook's men saw at a best few were to know again. Cook sensed that. So did the other thoughtful men on board. After the Friendly Isles, the ship passed south of the Fiji group (not seen since Tasman's voyage) and north-west again towards that other vague mark upon the map of the Pacific, Austrialia del Espiritu Santo, found by Quiros in 1606, to the delight of Dalrymple. Bougainville had passed through part of the group in 1768 but did not stay.

Here the Polynesian Pacific came to a halt: dark Melanesia took over. Perhaps there were then handed-down traditions of white men's murder in the islands Cook now named the New Hebrides. At Malekula there was open hostility: at Erromanga, treachery. Nonetheless Cook examined the group, found Quiros's Bay of St Philip and St James where he wished to found his ambitious New Jerusalem, sailed south to Tanna, surveyed and charted the group as well as possible. Not far away towards the south-west was the long island Cook named New Caledonia – a new discovery, populated by unique natives, both friendly and honest. Not a pilferer, not an instance of thievery was met. Fresh fruits and vegetables were abundant. The delighted Cook gave away breeding hogs and dogs, and wished he had time to stay longer: it was already the end of September, and he had yet another Antarctic and sub-Antarctic season to survive. Sailing south, towards the Isle of Pines, the ship got among the reefs again, in a sort of miniature Coral Sea. More heart-in-mouth, iron-nerved seamanship had to be used to get away. Sighting Norfolk Island – that frightful prison-island for the twice-condemned of Australia's convict era – Cook made for New Zealand and his base at Queen Charlotte's Sound. Now at last he knew that the *Adventure* had been there long before, and gone.

Now he could go home himself. It was early November 1774. There were four good summer months ahead, a little Pacific sector and some Atlantic sub-Antarctica still to investigate. The ship was overhauled as well as possible, the crew refreshed: on 11 November the *Resolution* sailed to run directly across the South Pacific on what was to become the New Zealand wool clippers' route on 55° South. Here again there was no sign of any land, but Dalrymple had a large piece on his chart to the eastward of Tierra del Fuego and Cape Horn, called the Gulf of San Sebastian. It was desirable to sail where that was, too.

First Cook did a very nervy thing, typical for him. He knew quite well the position of Cape Horn for he had himself fixed it: all he had to do now was get in a slightly lower latitude and run 5,000-odd miles before the west winds. This was the practice of all the square-rigged ships which sailed that way in their thousands long after him, for they all avoided the lee shores of southern Chile and Tierra del Fuego like the plague, for well-grounded fear of running on them. The difficulty was data for astronomical observations, for sun and stars down there were so often obscured. For this reason, and others, it made no sense at all for commercial sailing-ships to try to run for the Straits of Magellan. The *Resolution* was no commercial ship: Cook made directly for the coast by the western end of these dangerous straits. After a passage of thirty-six days, with constant sharp lookout and no driving, he was off Cape Deseado.

There followed weeks of careful examination and survey of that rocky lee shore, relieved by yet another Christmas celebration and followed by a stormy sail right across Dalrymple's last hope of *Terra Australis*. Nothing was there. Glimpses of South Georgia and the South Sandwich islands, both new discoveries, encouraged no one to look for prosperous lands in the area. Cook sailed carefully and sceptically eastwards near the parallel of 60° South, checking again on Bouvet's Cape Circumcision. The search yielded nothing. Near Good Hope he crossed his outward track of 1772, and his great southern west-to-east circumnavigation was completed.

On 21 March 1775 the good ship *Resolution* anchored at the Cape of Good Hope she had left in November 1772, almost three and a half years before, after a voyage such as never was made before, or since. He had discovered new groups and new islands in both Atlantic and Pacific, rediscovered the long-lost Marquesas and New Hebrides, pinned down the Antarctic, sailed Dalrymple's 'continent' off the sea for ever. And he had not lost a single man to scurvy or the bloody flux, or anything else avoidable. Three were drowned. One died from disease he brought aboard.

After five weeks at the Cape he sailed for England, with calls at St Helena and Ascension and time off to fix the position of the island of Fernando Noronha. Ninety-two days out from the Cape, three years and eighteen days after setting out on the voyage, Cook brought the *Resolution* back to England and dropped anchor off Spithead.

It was 30 July 1775. The War of the American Revolution had been raging for over three months. So there was no *Terra Australis*? It was unfortunate: for now it looked as if, soon enough, there would be no American Colonies, either.

The merits of that great and thorough voyage were overlooked in disappointment at its results.

CHAPTER TWELVE

North-West Passage

DESPITE the war, in which the old enemies France and (to some extent) Spain were to join, life ashore in the England of 1775 meandered along its accustomed eighteenth-century way. Wars were not so all-absorbing that some in the nation had no time to be interested in Cook's voyages. He had been at sea throughout six of the preceding seven years. He had made two tremendous voyages such as never had been made before. He had swept the map of the Pacific clear of the imaginings of the more perverse academics, pinned down Antarctica, defeated Cape Horn twice without even appearing to fight the place, established the sailing-routes to Australia and New Zealand for use when wanted, and – even more important – set up excellent relations with those cheerful but perhaps imperfect 'noble savages' of the South Seas, the Polynesians. He had filled in the blank faces both of New Zealand and Australia (either achievement would have been a sufficient life's work for another man, and beyond most). He had discovered or re-discovered almost every island-group of importance in the South Seas. He had brought his crews through shipwreck on reefs of extent and hazards not previously dreamt of: he had surveyed and charted with precision where his predecessors – if any – had so often been so vague and imprecise that they did not so much discover as indicate where discoveries might be searched for by better seamen: he had brought big crews of plain seamen, by no means specially selected nor even robust beyond the ordinary, in crowded little ships twice round the earth sailing in the aggregate over 120,000 miles, and he had lost no man to scurvy at all.

His achievements were immense, but felt by some as in the main negative and, by a few *Terra Australis* champions, perhaps even regrettable: there was an extraordinary reluctance then and for years afterwards to begin to realize the potential importance of both Australia and New Zealand, perhaps understandable on the score that both were so far away, apparently offering no immediate prospects of profitable trade, obviously without a cheap and useful native labour force or swiftly exploitable resources. If England now had to fight to subdue good but rebellious British cousins in the American colonies just across the North Atlantic, what might her prospects be with 'plantations' in the distant South Seas? Yet it was important to Britain that France and Spain should not become powerful there. Cook had at least forestalled these nations: sooner or later there must be a follow-up.

In the meantime there remained one last great unknown, not in the South Pacific but in the North. The chimera of a North-West Passage, guide way to China and the East around or through the north of North America – how about that? It had been sought, most valiantly and vainly, only through the ice wastes of the far North Atlantic. If there were any such Passage, obviously it must debouch in the North Pacific somewhere. Why not look for it from that side? Of course, this had been thought of before, and that efficient order-scorner the incurious Byron sent off to do just that. He did nothing whatever. But the indefatigable, inexhaustible Cook, the man who not only carried out his orders but used his judgement to better them, the man who held the confidence of his seniors and the trust of his juniors, the man who could keep inexpensive little ships at sea almost continuously for three years at a time and sail them safely to the ends of the earth – why, here was a seaman who would search thoroughly for the North-West Passage, survey it properly if found, and bring back the definitive answer whatever it might be. While he was about it, he might as well look for the North-East Passage around Siberia from the Pacific side too, in the absence of a useable route the other way.

What was required, of course, was not knowledge expensively acquired of any commercially useless, frozen sea which might – and in fact does – extend beyond the north of Arctic Canada and Alaska, but the discovery of some useable sea route ice-free and available, at least by summer, opening out from somewhere not too far from the north-west corner of Hudson's

Bay, flowing westwards towards the Great Slave Lake or Great Bear Lake and on to the ice-free waters of the North Pacific. No one then knew that all the Yukon, the mainland North West Territories of Canada, and all Alaska formed solid, united, and frequently mountainous land. Spanish seamen had some knowledge of the west coasts of the future U.S.A. and a little of British Columbia: in England such knowledge stopped at Drake's New Albion. Nobody quite knew where that was. Cook (if he went) would have to find Drake's Bay first, and start from there.

If he went? Of course he would go! His life's ambition seems to have been to leave no problems unsolved in the whole Pacific. He knew considerable satisfactions from that second voyage: at last he was promoted post-captain, a notable achievement for the ex-Mate of a Whitby cat. But, one would think, long overdue. Again he was presented to George III. He read outstanding papers before the Fellows of the Royal Society, one on the tides on the east coast of New Holland, the other on the preservation of the health of long-voyage seamen. For the second, he was awarded the Society's Copley Gold Medal. He was elected unanimously a Fellow, the highest mark of approbation for scientific and philosophic achievement in England and ranking high in all Europe – an honour seldom offered the non-academic or the self-educated, as Cook was. He dined out with some of the great and the near-great. He found himself acclaimed among the intelligentsia in France, in Spain, in Russia and in Holland as one of the great discoverers. He had the satisfaction, too, of proving finally and decisively that the Harrison chronometer was the answer to the longitude problem, for the copy of Harrison's instrument – the so-called Kendall's watch – carried in the *Resolution* worked splendidly throughout the voyage during which it lost a little over two minutes a year, at a regular rate.

The Harrison chronometer had been well tested long before this from 1736 onwards, but the jaundiced Board of Longitude – committed to the Lunars system – long refused old John Harrison his award. There was then a great deal to be said for longitude by lunars once nautical tables and an accurate sextant were available, particularly as the early chronometers were extremely expensive, each being made by hand over periods of years. The publication of the Nautical Almanac in 1768 – Cook carried a copy of the first in the *Endeavour* – simplified the working of lunar problems, but they were still difficult. The award, £20,000, was for the perfection of a practical and dependable means of discovering longitude: this Harrison had done. The old man was dead by the time Cook was back from his second voyage, and the reward had at last been paid. Cook was loud in his praises: the life-time's efforts of the old Yorkshire carpenter's son were thoroughly vindicated. Not that that made much real difference. It was over another quarter of a century before Admiralty gave general approval for chronometers to be issued to H.M. ships.

Within a day or two's sail of St Helena, homeward bound on the second voyage, Cook spoke the *Dutton*, East Indiaman. There was the customary communication by Cape Horn voice, quarter-deck to quarter-deck across a cable or so of the gently rolling sea.

'What do you make the longitude?' shouted the East Indiaman.

Cook told him. They laughed.

'You'll never make St Helena! You'll miss the place by a hundred miles!' shouted the *Dutton*'s captain.

'We'll see,' said Cook. 'Follow me, and I'll put your jibboom ashore there if you want!'

In the morning, the blue smudge of St Helena was in line with the *Resolution*'s bowsprit-end and the *Dutton* was nowhere to be seen. Now Cook laughed: but his laughter (a cheerfulness not often reported) died down when he went ashore, for the ladies there had read Hawkesworth's (alias Cook's) official book and were extremely cross, for the ridiculous Hawkesworth had taken some ill-considered criticisms of life at St Helena from Banks's journal and put them into Cook's mouth.

'What *do* you mean, Captain?' asked the ladies very courteously, but all their eyes flashing. 'We are cruel to our slaves? We have no wheeled vehicles? We don't even know about the wheelbarrow? Pray explain, if you please!' And so forth.

In vain the captain, faced so bluntly with the misdeeds of an aspect of the writing trade new to him, explained that he had never written such stuff, nor read it, until a copy reached him at the Cape: the author had access to other journals beside his.

The ladies did not believe him. Their satirical sallies, remarks Anders Sparrman, 'probably troubled the naval officers more than any storm'. And, every morning during his stay ashore, the captain found most of the wheeled vehicles and all the wheelbarrows in the island carefully parked outside his door. As he came out, a group of smiling slaves gave him good morning in their sonorous, perfect English.

Cook privately cursed scientists, and Hawkesworth too, but the writer had died two years before the *Resolution* returned. This time Cook could at least work on his own book, though a literary Canon of Windsor named John Douglas – a pleasant, able, and really helpful man – was retained to give any help he could.

The literary side of the second voyage was effectively spoiled by the Forsters, when the son George rushed out a book on behalf of them both – much of it well-written, some of it moving, but far too high a proportion ill-considered, captious and complaining; some of it downright libellous.

'Damn and blast all scientists!' Cook is reported to have exploded one day in the *Resolution* later, when Lieutenant King called on him. The lieutenant had just been appointed to the ship for Cook's third voyage, and, perhaps foolishly, ventured the observation that it was a pity no scientists were to come this time.

'Curse the lot!' said the great seaman, a little short-tempered that morning, and sour on shipboard scientists, cluttering up the ships, making excessive demands, never once getting into their thick skulls that the primary care of all shipmasters must be the ship herself and the safe prosecution of her voyage. And never appreciating the wear and tear, the endless searing stress, of the world-wandering wind-ship on the lonely man who sailed her – lonely indeed, though his great cabin might be filled with the arrogant and the self-important.

Cook accepted command of the third voyage, of course. He

would finish the job. He was forty-seven years old, had been at sea more or less continuously for thirty years, deserved a real leave, if not retirement. His friends at Admiralty appreciated this. They had appointed him to a captain's berth at Greenwich Hospital, a much-sought sinecure meant to assure the recipient of a good place with reasonable income and no excess of duties for the rest of his life. Here at last Cook might have sat back a while: but he accepted the appointment only on the understanding that he could leave at will if there were ever greater need of his services. He knew very well what form that greater need might take: but there was a regrettable haste about beginning the fateful third voyage. Cook should have had quiet rest for at least two years and a thorough medical check at the best hospital there was: that appalling upset in his stomach to which Forster Senior sacrificed his dog, down in the South Pacific, had never thoroughly cleared up. What, if any, deep-seated ills the captain might be suffering no one knew or bothered to find out, including himself. It appears that he kept all news of his intention to make another voyage from Mrs Cook for as long as possible. Poor Mrs Cook! Another child had died: two surviving boys were already in the Navy: she dined in the homes of the great with her famous husband on occasion, but apparently not often: she must have been delighted with his Greenwich appointment for so many reasons – above all, it meant no more sea! And he could live at home.

No more sea? She did not know of his secret promise.

On 13 September, when the ship had been back in England less than six weeks, an Admiralty minute ordered that the *Resolution* be put forthwith into condition for another 'voyage to remote parts' and the Board informed when she would be ready to receive men.

It took a long time to refit the sloop. There was heavy pressure on the dockyards to provide shipping for the American war – transports, blockading frigates, ships-of-the-line, sloops, everything. Small sloops for 'remote parts' had low priority. Suffering, too, from some maladministration and spread of dishonest practices with the great increase of opportunity for profit from them, the standard of work in naval dockyards was often far from good and sometimes poor indeed. The *Resolution* got a hurry-up refit, badly skimped. Cook had brought her home in good shape, considering what she had been through. Now she was poorly sparred, indifferently caulked, weakly rigged with far from first-class cordage. But when the time came, she was well manned and stored, and – Cook certainly appreciated this – on this voyage he was to be his own astronomer and 'scientist', and Surgeon William Anderson would be botanist and naturalist to the expedition. Anderson had made the second voyage, was an excellent physician, surgeon, botanist, and linguist and a good and well tried shipmate. He was worth the two Forsters put together with Sparrman added: the absence of the Forsters was a tremendous relief. (As for Forster Sr, his impossible cantankerousness soon disgusted everybody: he had quarrelled with Admiralty and every other body or person with whom he came in contact, and soon was on his disgruntled way back to the Continent.) The painter Webber was to be official artist – the eighteenth-century equivalent, perhaps, of a skilful photographer – and the Polynesian Omai, who had come home with Furneaux and been a great success in London, was to travel back to the Society Islands. Executive Officer of the *Resolution* was John Gore, who had made the *Endeavour* voyage: Master was a fine seaman named William Bligh who was to make several voyages, at least one unpremeditated: the crew included six midshipmen, a cook and a cook's mate, six quartermasters, twenty marines including a lieutenant, and forty-five able seamen.

Furneaux's *Adventure* had gone off on another commission. She was replaced by another Whitby collier, the *Discovery*, of just under 300 tons. The *Discovery* was commanded by Charles Clerke who had been an officer in both the *Endeavour* and the *Resolution*, with James Burney as his Number One. Lieutenant Burney had begun in the *Resolution* on her previous voyage as a midshipman but was soon promoted and transferred to the *Adventure*. Between the *Adventure* and the *Discovery*, Lieutenant Burney had managed a year's service in the frigate *Cerberus* on the North American station.

Both the *Resolution* and the *Discovery* had a useful leavening of experienced circumnavigators. Ramsay the cook stepped up to gunner's mate of Cook's ship – an odd change of duties though doubtless to his considerable satisfaction, for guns were easier to serve than mariners. William Ewin, American bo's'n's mate (from Pennsylvania) with Cook on the second voyage, now let himself out of his countrymen's war – as so many seamen did, most of them unable to help themselves, for they had often to serve the first ship at hand, English or American – and signed as boatswain of the *Resolution*. He was not the only American aboard. Another was a corporal of marines by the name of John Ledyard or Lediard, a New Englander of good family but, unfortunately, an unstable character who was to cause some trouble with ill-considered accounts rushed into print, after Cook's death. The East Anglian George Vancouver, then aged nineteen, was a midshipman in the *Discovery* with his friend Richard Hergest – another from that voyage who was later to be murdered by the hot-headed Hawaiians, at Oahu in 1792 – out of the *Daedalus*.

While new crew were being assembled, old were not forgotten. The widow of Murduck Mahony, late departed cook of the *Adventure*, was found a berth as nurse at Greenwich Hospital. Cook wrote specially to his friend Lord Sandwich for her. Her job was no sinecure.

On 12 July 1776, almost a year from the date of return from his second voyage, Cook took the *Resolution* to sea from Plymouth, that Devon port from which so many of the sea's great had begun their voyages. Her decks were like a cattle ship's with livestock sent by George III as gifts for the South Sea islanders, as well as the usual bullock or two, a small mob of sheep, the odd she-goat and a few hogs and poultry for the ship's fresh food.

'All we need now is a few females of our own species,' said Captain Cook, very grateful that no such were there.

At Plymouth the Sound was filled with ships bound to the westward for the War of the Revolution. This made much difference to Cook only when France came in, for there was little real American Navy then to stop him, if it wished. The enlightened Benjamin Franklin, then Minister Plenipotentiary from the Congress of the United States at the court of France, was already seeking safe-conduct for him – a plan which an unenlightened Congress, with much on its hands, tried to rescind when it heard of it. Congress knew that the chances of intercepting Cook (if anyone wished) were slim, anyway. After Franklin's example, France announced a similar safe-conduct when she came into the war.

The war saddened Cook. A few days before he sailed, he watched 'his Majesty's ships *Diamond*, *Ambuscade* and *Unicorn*, with a fleet of transports consisting of 62 sail, bound for America with the last division of the Hessian troops and some horse, which were forced into the Sound by a strong north-west wind'. They made a magnificent sight as the three-score ships beat in led by the great three-deckers, with the roar of the wind in all their riggings, the red-coat's splurges of brilliant colour everywhere, the war horses neighing on deck. To Cook it was 'a singular and affecting circumstance'. He regretted the 'unhappy necessity of employing H.M. ships' for such a purpose: and he was not alone.

There were a few horses aboard the *Resolution* too, bound neither for the rebellious American colonies nor as replacements for their slaughtered kind already in the salt-horse casks in the ship's hold. If they survived the voyage, they were meant to impress the Islands' potentates and to be left with them as gifts.

The congressmen in Philadelphia were hardly to be blamed for thinking the voyage a little odd. It was a strange time to seek new discoveries on the west coast of North America when the east coast was battling to break its ties with those same discoverers. It was true that no one but a few brave Spanish seamen (and, far to the north, an isolated Russian fur trader or two) knew anything as yet of the north-west coasts of North America. Cook's orders were to sail first to the South Indian Ocean to check there on certain discoveries made by the French – obviously to gauge their value as possible naval bases in the struggle for power in the Indian Ocean – then to land Omai at Tahiti or his own island. After that, he was to sail for the North Pacific and the coast of northern California. There he was to examine everything northwards of Drake's Bay until he found a sea passage through to the Atlantic, or that there was none. This would take some time – at least two summers, if it meant (as it did) more ice navigation. Bases for wintering were to be sought in Kamchatka or elsewhere as discoveries and conditions might indicate. When he found the North-West Passage (or, perhaps more doubtfully, the North-East) he was if possible to sail back to the Atlantic by that means, making careful survey as he went.

Cook and his officers and men were on their ordinary naval pay, but a national prize of £20,000 had been voted by parliament in 1745 for any British ship, not the King's, which found a useful North-West Passage out of Hudson's Bay. Twenty years later, land and sea exploration had put an end to hopes of any such discovery from Hudson's Bay. The act was now amended to include H.M. ships (and Cook), and any North-West Passage would qualify so long as it was beyond the parallel of 52° N. There is no evidence that the 'prospects' of £20,000 made any difference to Cook. There had been the Board of Longitude's £20,000 too: and old John Harrison had an indisputable claim to that for the best part of half-a-century before he received it. The prize, however, was a considerable encouragement to the mariners. They felt that there were two advantages, at any rate, in going with Cook. If the Passage *was* there he would find it: and he would bring them back alive.

But indeed the whole projected voyage was, to say the least, somewhat optimistic. It added up to the toughest, longest voyage that even Cook had taken on – and the most hopeless too, for that matter. The very business of inducing the sea winds to move two small sailing-ships in safety over that immense waste of waters would be daunting, even now. From the English Channel in 50° North, right down the lengths of both Atlantics, round Good Hope and through the boisterous rush of the wild west winds for 10,000 miles, over almost the entire length and the whole breadth of the Pacific Ocean, from sub-Antarctic to the Arctic and – having survived all this, with calls at all sorts of islands and continents, friendly and unfriendly, dangerous or paradisial – back to England by any means that offered. It meant rounding Good Hope and probably the Horn as well, crossing the equator twice in the Atlantic and at least twice in the Pacific. It meant months of dangerous sailing among both ice and coral reefs. It was a voyage, one would think, to strike dismay in its very contemplation even in the lion heart of Captain Cook. All those long sea miles to be so slowly sailed, each so hard to gain! And tracks so painfully pioneered crossed and re-crossed, and crossed again, almost without end.

It was perhaps also an unnecessary voyage. What real chance could even the politicians think there was for the existence of a useful North-West Passage, or a North-East? North Polar seas had been probed often enough and the questing ships balked by ice always. It was an interesting idea to look, but unwarranted optimism to seek the Pacific exit of a channel where Atlantic entrance was unknown and probably non-existent. What could the courageous and most competent Cook do to the ice that it should open up for him? Tradition of failure in that quest dated to the pre-Columbian era: did each generation have to taste freezing hopelessness for itself?

But the rewards of success could be great. So, for that matter, could Eldorado's.

After her first brush with an Atlantic blow, the *Resolution* leaked abominably. The mariners hanging at the bow could see the caulkers' ill-done work weep from her seams as she lifted and plunged. Badly caulked decks worked more than the ship did and the resultant openings let in the cold seas that washed over her. All the quarters were wet and the spare sails became sodden and mouldy in their lockers. Water seeped down the ceiling in the 'tween decks and into the store rooms, not seriously, but miserably and destructively. The poor-quality hemp shrouds, badly rigged, were set up hurriedly again – not a good idea with the ship working in a sea-way, for the setting up of shrouds and backstays to support wooden masts was a fine skill in which an error could strain a masthead and cause serious damage, latent at first perhaps, breaking out at awkward moments. They aired the sails, crowding the livestock to find room to spread the wet canvas. They opened the hatches, rigged the wind-sails (which were chutes), set the charcoal fires going in their moveable pots whenever the ship's motion allowed. Leaks in the forefoot and elsewhere made pumping necessary each watch. This was not serious but it was annoying, especially as they had brought the ship home in such a good state.

Perhaps Cook blamed himself, at least in part. A ship in dockyard hands has to be looked after even more carefully than on a voyage, for the indifferent workman can do her infinite harm, cover it and go away. With his Greenwich Hospital appointment – kept on as long as possible for the sake of Mrs Cook – his reports, charts, literary work, social and Royal Society life, and his rank of Captain, it was not easy for Cook to get from his home to the dockyard. Since, too, Mrs Cook did not know of his plans until not long before the *Resolution* sailed, it is probable that her husband could not take too obvious a personal interest in the ship throughout her refit – a general interest, perhaps: that is not the same thing. Anyway, here she was, setting out on so tough a voyage, a leaking, ill-rigged sieve. There were plenty of good practical seamen aboard to cope, from Cook and Bligh on down: but the sea is not a dockyard. There is evidence that on this last voyage Cook was not always his former self. Outbursts of temper were perhaps a little more frequent and actions in them sometimes ill-judged – new for him; but understandable. How troublesome was that malign and so strained digestive system? How tortured that iron will? We don't know. He doesn't say. Mrs Cook didn't say, either: or if she did, nothing was recorded.

Before the Cape the mizzen-topmast was found to be sprung – cracked so seriously as not to be able to bear sail. The mizzen-topmast was not at all as vital as the main or fore topmasts were, for the chief purpose of the mizzen was to carry the fore-and-aft spanker (set to the lower-mast) and provide useful leads for the main tops'l and t'gallant braces. But a ship on a long voyage needs *all* her masts. Which might be the next to go?

At the Cape Cook bought a replacement, and set the crew to re-caulking overside and on deck, along the waterways, in the eyes of her, down the forefoot, round the stern-windows and the counter, everywhere the carpenter's large gang could get at. Now the work was thoroughly done. In places overside they

found the oakum spewed right out of the seams as if it had been put there by fingers. Hard winds at the Cape and a considerable sea running at the anchorage were no help. Fresh-baked biscuit replaced some lost on the voyage: the livestock were landed to graze and recover from the sea passage (and dogs destroyed the best sheep): the Kendall-Harrison chronometer – once more aboard and behaving splendidly – was checked for its rate by observations (for there were no time signals): the crew fed on fresh food including plenty of greens daily. A sick member of the crew, much bothered by the wet quarters, was sent home by Indiaman, and as much new livestock as possible bought both for future gifts and provisions. This included four horses, rabbits, and poultry. The *Discovery*, which had sailed out to the Cape independently, arrived late and had to be recaulked, too.

By 30 November 1776, both ships were ready and that day they sailed. It was Cook's last contact with civilization. Away he sailed with the wild west winds to 45° S. and almost to 50° S., and surveying the French discoveries of Marion, the Corzet group, and Kerguelen on the way east. These were rough places on or just inside the ice line – the northern limit of Antarctic drift-ice – and somewhat unlikely ever to have much real importance.

On sailed the little fleet across the bottom of the world towards Tasmania and Cook's favourite Queen Charlotte's Sound in New Zealand. In a shift of wind flung at the ship by a vicious squall, the mizzen-topmast came down though no sail was set on it. It went clear of the decks and no one was hurt. Again in a sudden violent squall at four o'clock in the morning of 19 January, in about 45° S., the fore topmast went by the board and brought down the main topgallantmast with it. This was a mess, a really serious partial dismasting. But the masts did not go overboard for their own rigging clung to them. It took the whole day to clear up, saving everything possible – the rigging, sails, and spars. There was a spare fore topmast on deck. This was sent up and rigged. There was no spare main topgallantmast so the rescued fore was fished and shifted to the main, leaving the ship with a full mainmast, a mizzen lower-mast, and a bit more than half a foremast – quite enough in the Roaring Forties, but the shortened rig gave her an odd look. All this carpentry and rigging aloft was herculean work, with the ship rolling violently throughout as she ran before the gale.

Cook diverted towards Adventure Bay, the Furneaux anchorage on the east coast of Bruni island off south-east Tasmania, to find trees for new masts and fresh fodder for the unhappy livestock. It was his only visit to Tasmania, a beautiful island with excellent harbours and some of the best ship-building timber in the world. Adventure Bay is not far from Storm Bay which leads to the estuary of the Derwent River, one of the finest harbours in the southern hemisphere if not the world. It is a pity that Cook did not sail on a little and discover this, but it seemed his destiny to miss the best harbours. He was in a hurry, hoping to catch at least something of the northern summer along the coasts of North America. It was already nearing the end of January. His needs were few – spars, fodder, firewood, fresh water. He knew all these were to be had at Ship Cove, in Queen Charlotte's Sound, New Zealand, and the Maoris there were friendly and spoke Polynesian. So he dropped anchor where Furneaux reported anchorage, met Tasmanian aboriginals who were friendly but primitive, caught an abundance of fish, cut a few spars, as much grass as he could, and some firewood, hove up and sailed on. It proved difficult to find any point of contact with the few Tasmanian natives met, which was a pity. The unfortunate Tasmanians were all gone a century later. Cook did note that they had no canoes, not even a satisfactory sort of a dugout or bent bark canoe for beach fishing. He took this as evidence that they must have come overland, supporting the Furneaux view that Tasmania was joined to and part of New Holland.

A passage of ten windy days brought the ships within sight of New Zealand. At Ship Cove, Cook met the murderers of the *Adventure*'s boat crew. As that crime was three years past and unpremeditated – a sudden flare-up in which the English seamen shot first and the consequent rush of the Maoris overpowered them – Cook made it clear that he had not returned for vengeance. To the surprise of the natives, amicable relations went on exactly as when Cook had been there before, to the disgust of Omai the Tahitian. He urged summary justice for the consumers of the English seamen, declaring that it was right and expected, and indeed some of the Maoris were asking what kind of men were these who did not avenge the blood of their fellows?

From the contemporary Polynesian viewpoint Omai might have been right. Cook paid no attention: to these people sudden death was commonplace and acceptable as the sought-after end in a quarrel. They meant little by it and it meant little to them: the quality of mercy was new, and this was Cook's contribution. Omai said they scorned it: but Cook had long suspected that Omai was rather a dull young man, conceited from his stay in London.

Delay for spars and livestock fodder was rapidly destroying the chances of exploring in the far north that summer. Drake's New Albion was a long way off over seas quite unknown (except for the tracks of the old Spanish treasure galleons) once Cook passed the Marquesas. It would take the whole of the 1777 summer, probably, to reach Alaskan waters: what then? More difficult ice navigation, as in the Antarctic? Pretty likely. That would take time, too. It was near the end of February before the ships got away from New Zealand. This time Cook made a north-easterly course which brought him to the Hervey Islands (now the Cook group) – another new discovery, which offered poor or no anchorages and little refreshment. Again the crowded livestock were in serious want of fodder: Cook now accepted that he could not reach North America that summer and made for the nearby Friendly Islands.

Here he was assured of a good reception and most of his needs: but thievery was worse than ever and so were navigational difficulties. Friendly chiefs themselves were not above active robbery: Cook caught one, fined him a fat hog, and had him given a dozen lashes which he accepted stoically as fairly due. But he still stole, and so did his people: they seemed to steal in spite of themselves. Captain Clerke of the *Discovery* thought of the idea of shaving the thieves' heads and for a while this did some good. They hated to lose their long locks and they disliked the ridicule too: but audacious, bare-faced thievery went right on. On one persistent offender Cook lost his temper and had his ears cropped, as well as his hair: his officers looked at him with some worry. This was not like their Captain Cook, though in this case the thief was an old offender, and he had stolen that vital instrument, a sextant.

Sailing among the islands both ships struck outcrops of coral, fortunately very lightly. Even more fortunately, it was loose coral. They came off easily, without damage. Again the *Resolution* sailed over a clear sea towards an unsuspected, sudden sandbank where her complete wreck seemed imminent. The bank was not a cable away when the discoloration indicating it was seen: but Cook and his crew were ready. The watch was standing by at stations for tacking ship, the Master, Mr Bligh, in charge of them.

'Down jibs! Leggo bowlines! Spanker to wind'ard! Lee-oh!'

The well-known orders, roared suddenly from the quarterdeck, set in train an instant rush of the bare-foot seamen, a thrashing of canvas, a rattle of reef-points, a swift spinning of the wheel, a brawny-backed hauling of braces, shifting of bow-lines, tacks and sheets, lifting of course clews, swinging of yards, setting of sails again. She swung in her stride from tack to tack without losing headway: the bank stood a moment, a menacing trap in her path: she pulled back from it faster than any liner may go astern: flapped her sails and was off. But how many more banks? How often could she hope to get away with this sort of thing? Constant, hundred-per-cent vigilance was the answer, day and night. It was a demanding standard, hard to maintain, hard on the Captain too.

Life was not all difficult sailing or dealing with hopeless thieves. There were entertainments, native theatre, dancing, displays by the Polynesians; fireworks and the marines going through their drill, organized by Cook. At Tongatabu the paramount chief or king gave Cook a special house ashore: gifts of fat hogs, poultry, coconuts, fruit and greenstuffs were abundant. Cook distributed King George's livestock – rams and ewes, a bull and two cows, a horse and mare, English goats, a fine English boar and sows, buck and doe rabbits: not forgetting gardens of useful vegetable and fruit seed. Even here there was more thievery, bare-faced and insolent. Cook seized canoes, houses, and hostages among the chiefs. These came willingly, for they knew Cook. Their incarceration lasted only until the stolen items were restored, which usually did not take long. Most probably the chiefs had instigated the thievery. All this became a sort of a regular drill, an essential part of any South Seas visit – tumultuous, warm and completely friendly reception, games, dances, fireworks, drill, trade, thievery, seizing of hostages, return of stolen goods, release of hostages, more thievery, more hostages and so on *ad infinitum*.

After the Friendly Islands, Cook sailed for the Societies and Tahiti. These were to wind'ard in the trade winds: reaching them meant more delays, for the trade wind was light. Omai had to be landed: Tahitian, Huahine, Raiatean and Bora Boran chiefs visited: good relations maintained, and the lesson that there could be harmony and lasting good from all these visits brought home to the islanders. Hence Cook's more or less regular visits. Again the routine was established. Two Spanish ships had been at Tahiti while Cook was away: it was pleasing to note that the Spanish seamen had acted in accord with the Cook standards – no clashes, no disharmony, nothing but the most pleasant memories (and the present of a noble bull) were left by them. Cook shared out all the remaining livestock – three cows for the bull, another horse and mare, sheep, a turkey cock and hen, a gander and geese, a drake and ducks, even a pair of breeding peacocks. (The safe carriage of all these over so great a distance – most were from England – was a considerable accomplishment: the open decks of the *Resolution* and *Discovery* must have been hard on livestock, and so were the Roaring Forties.) All hands breathed better (and cleaner air) when the farm had gone.

Then, as sailing time approached, two of the crew deserted, a midshipman and a seaman from the *Discovery*. The remarkable thing was that many more did not try from both ships. They were invited: what had the seamen to lose? Cook knew that successful desertion might start an exodus. It could not be allowed: but at first the deserters could not be found. Cook seized canoes, houses, chiefs, made shows of strength, demanded that his men be brought back no matter where they were, for they could remain ashore only by chiefs' connivance. To show that he meant action, Cook himself took armed boats to search for his men. Rigorous steps brought swift action: the deserters were picked up on Bora Bora and returned.

In the meantime Captain Cook made an interesting discovery in yet another field. He was suffering badly from most painful, crippling rheumatism, especially from the hips to the foot, intensified by the wet quarters caused by the ship's appalling refit.

'We'll fix that,' said a friendly chief. Surgeon Anderson – a very busy man in the islands – approved the idea on the grounds that it could do no harm and might do good.

So twelve large, muscular women, the chief's relatives, were paddled out ceremoniously in a large canoe and, being properly received, descended to the great cabin. A mattress and blankets were spread on the wooden deck.

'Lie,' said the women.

Cook lay down. The twelve large women immediately fell

upon him, pummelling, squeezing, and massaging unmercifully with their plump, lively hands from head to foot but especially the rheumatic joints, until his bones cracked and all his flesh looked and felt like misused blubber. For fifteen minutes that seemed like hours the determined ladies continued these ministrations, cheerfully and with both hands. At last the released victim got up. To his astonishment he felt immediate relief.

'More?' asked the ladies, smiling.

Indeed, agreed the Captain: two more treatments cured him. The rheumatism, pummelled so skilfully and mercilessly out of him, went away and did not return.

Now it was time to leave familiar places and stand north. The sailing-ship way to reach California from the Islands was to sail first directly north with the south-east trade wind on the starboard beam, work through the Doldrums belt the best way possible and then, picking up the north-east trades also on the starboard beam, sail north out of them. It is futile for square-rigged ships to beat against trade winds unless compelled: far better to use them to make latitude and then, with the variables and westerly winds beyond, run eastwards. Cook was pioneering this route, which was to become greatly used later: he knew the wind system well enough though he had not been before in the North Pacific. It was much the same as that of the South Pacific – if anything, slightly more reliable. The two little square-riggers kept nicely together on the way towards the perhaps ambitiously named New Albion. Drake had coasted there and had been rather vague about his discovery.

On Christmas Eve, 1777, Cook sighted the island he named Christmas, which is now one of the Line Islands. It was mid-winter in the northern hemisphere: he caught fish and turtles, and pressed on. Sixteen days later, near the limit of the trade wind, a large, high island showed ahead – then another, and yet another, all three high, bold, picturesque, volcanic land, not coral atolls.

It was 18 January 1778, a beautiful day with the white-caps talking back cheerily to the bow-waves as the two little ships hurried fussily along, a wonderful day of perfect trade wind sailing across the azure sea. Cook shaped his course for the second

island. It was beautiful but poorly supplied with harbours. He managed to get an anchor down: canoes came out manned by mild fellows, without arms, who spoke Tahitian. How had the Tahitians sailed over so huge an area of the Pacific? Cook and his seamen wondered, for now they had encountered the same people over a fantastic area – from Easter Island to this place called by the natives Atui (now Kauai, in the Hawaiian Islands), from Tongatabu to Tahiti, from Bora Bora to New Zealand. Cook gazed at them with heightened interest. He had one or two tolerable interpreters in a sort of pidgin-Tahitian among his veterans, and many more who thought they were: he could make himself understood in the South Seas *lingua franca*, and follow what was said to him. When anchorage was found, the islanders brought pigs, potatoes, plantains, sugar-canes, parting with these quietly for whatever was offered them and darting their bright eyes everywhere about the ships like men who had seen nothing like this ever and could scarcely believe it now.

Unfortunately, many began at once to take whatever items about the deck caught their fancy and were removeable – not stealing, but simply lifting, removing, as wonderful things brought from another world for them by these tall sailing-gods. The sailing-gods, however, were not at all pleased and took all the ship's gear back again, with some degree of expostulation. The islanders seemed to follow that all these strange and attractive items were essential to the sailing of the ship (or island, as they thought it): when they grasped this they took no more. Except for one daring fellow who, quite unable to resist such a treasure, suddenly grabbed a butcher's cleaver and, leaping overboard with it, swam quickly ashore.

When Cook landed all the natives in sight fell flat upon their faces 'and remained in that very humble posture till, by expressive signs,' wrote the captain, 'I prevailed upon them to rise.' A long speech (or prayer) was offered, presents exchanged, friendship pledged, a brisk trade set up in excellent roasting hogs and other welcome fresh food. As he moved about, never very far from the beach, his approach was announced and everyone, wrote Cook, 'fell prostrate on the ground and remained in that position until we had passed'. The practice puzzled Cook, until he gathered that it was only 'their mode of paying respect to their own great chiefs'.

Perhaps. It dawned on none of the seamen that it could be something more than that.

Supplied with a sea-stock of fresh food, and well satisfied with the visit to the pleasant islanders, Cook weighed and stood again towards the north, seeking the expected west winds to blow him to New Albion. On 40° North, early in February 1778, the west winds found the ships; a month or so later the coastline of a great continent came in sight – the west coast of North America. It was a great and beautiful land with distant snow-topped mountains, very grand and high. But the westerly winds were on-shore, putting the ships in the inevitable danger of all square-riggers on a lee shore.

Cook prudently came up to the wind well out at sea, and stood towards the north, to make a running survey of the land and approach closely whenever opportunity offered.

CHAPTER THIRTEEN

Death in Hawaii

THE discovery of a sailing route across what now is Canada and the north-western United States was obviously impossible. Cook could see that great mountains blocked the way: while these rose majestic in the background it could hardly matter much what bays, inlets, gulfs or river-mouths might be found in the foreground. Unlike theorizers ashore, he was well aware of the real immensity of the land area where this chimerical open-water passage was imagined to be. He had surveyed much of the other side, the Newfoundland coast, the Gulf of St Lawrence and all that area. Now he knew the longitude of the west coast of the Americas, too (and in this was also pioneering). What a mighty country! From Cape Race in Newfoundland to the coast of Oregon, 70° of longitude, over 4,000 miles! As he sailed slowly northwards the coast trended towards the west – always to the west, sometimes slightly, often boldy. The wind was west, forcing him in towards the land: there were fogs, gales, appalling squalls. This coast of the states of Oregon and Washington was to become notorious in the sailing-ship era: Cape Flattery (named by Cook) at its north-west extremity rated then (and now) with Hatteras, the Horn, and Good Hope as the four most vicious headlands in the world.

On sailed the little vessels, feeble, man-made, wind-blown chips upon a hostile immensity, held to their course only by the ability and iron will of the great seaman commanding them – not just leading them, hidalgo-like, with discourse and majority verdicts of ships' councils of the real seamen frequently convened: actively *sailing* one, ordering the other, dominating everything and everybody.

Cook took the ships into Nootka Sound on the western side of Vancouver Island (so named much later by Vancouver himself) where the natives traded fish and furs and practised thievery on a professional basis. Sailing from here towards the end of April 1778, the *Resolution* opened up alarmingly while lying to (under fores'l and mizzen stays'l, says Cook) in a severe storm: in the bread room below, the sea could be both heard and seen rushing into the ship. Pump and bale was now the order, while the seamen worked through the long night of screaming wind and violently tossing ships. Even the taciturn, much-forgiving Cook must have reflected harshly upon the infernal shipwrights in the naval dockyard who had done their work so badly. The leaks, for the moment, got no worse: they were on or near the waterline, and, when the weather eased, could be dealt with. But more leaks developed. It was in rather an unnecessarily handicapped state that the *Resolution* – good ship as she should have been – approached the Arctic. Cook had to find a safe and convenient spot at least to list the ship and fix her increasing leaks.

He was now well inside Alaskan waters and had already sailed past the sheltered, beautiful inland passage of bays, sounds, channels, and arms of the sea which reached from Puget Sound, inside Vancouver Island, towards Prince Rupert, Ketchikan, Juneau (now the Alaskan state capital) and points north. He kept well out to sea. This was a north-south passage, along the coast, sheltered by off-shore islands, lapping the bases of picturesque and often precipitous mountains, fed by their glaciers – a gloriously beautiful place, heaven in summer for small vessels, but futile as a lead towards any hope of a North-West Passage. Off this waterway innumerable sounds, steep-sided gulfs, fjords and sea-drowned valleys opened into the mountainous hinterland of British Columbia and south-west Alaska. It was just as well that Cook missed the lot. They would have wasted time and led him nowhere. It was a pity, perhaps, that he also missed Valdez, at the head of the Valdez arm off Prince William Sound, for this was a good place though cold enough in winter. It was ice-free the year round, though north of 60° N. Cook was in the vicinity for some time, scattering good English naval names like Cape Suckling (Nelson's uncle), Hinchinbrook and Montagu islands (or Sandwich), Bligh Island.

From this area the Alaskan mainland turned southwards, to Cook's surprise. Before turning with it, he groped about looking for a north-going channel and still in search of a place if not to careen at least to list the ship. There was no way of avoiding a swing to the south, in due course, and fog, hazy weather with rain squalls and fresh wind made finding a suitable harbour difficult. One was found off Prince William Sound, not far from Valdez in a sheltered spot which Cook called Snug Corner Bay, off his Sandwich Sound, north of Montagu Island. Listing the ship at anchor – a much simpler process than careening – showed the oakum all gone from some of the seams below the wooden sheathing. This was made good (with, doubtless, another curse from the mariners for the criminal work of the shipwrights), while the local Eskimoes came out in kayaks and skin canoes. They obviously regarded the ships as gifts for plunder, but an attempt to seize the *Discovery* by a large force of them armed with knives was routed by the sailors with their cutlasses. Cook used no fire-arms: he wanted no murder.

Observations showed the ships still to be over 1,500 miles west of any part of Hudson's or Baffin's bays. Despite the appearance of many arms of the sea leading off the Sound, this and the behaviour of the tides showed that it was a waste of time to seek a North-West Passage there. It was already May. Cook must push on north: to do this he must first go south-west and west. Mr Bligh and Lieutenant Gore looked by boat into the more promising gulfs, but they were all useless.

The *Resolution* and *Discovery*, sailing very carefully, had not gone far to the south-west before coming to a headland round which the sea swept again to the north-north-east in a broad gulf which stretched away as far as Cook could see. Could this be it, the long-sought passage through? The North-West Passage? It was at any rate the best lead yet. Cook was delighted, as he swung round to sail into this 'Gulf of Good Hope', as he charted it. North-West Passage or not, this was obviously a considerable discovery, a fit place for famous names. Capes Banks and Mulgrave were the south-western points with Port Sandwich between them: King George's Foreland marked the extremity of the 'Land of Good Prospect' along the starboard hand.

Most of these names failed to survive. So did the hopes of a North-West Passage, for the gulf though deep was fed only by rivers tumbling down from the wildly beautiful hinterland. Its waters became brackish, then shallow and almost fresh – fatal signs that no sea reached them from north or east or, indeed, from anywhere. Mr Bligh the Master checked by boat: indeed there were 'abundant proofs' that they were in a river – 'low shores, very thick and muddy water, large trees and all manner of dirt and rubbish floating up and down with the tide. ...' The 'Gulf of Good Hope' became the river of disappointment: today it is Cook Inlet, a famous reach of broad waters, mountain-lined and strikingly beautiful, leading to the city of Anchorage and the great airport where the big trans-Polar aircraft put down briefly on their long flights between north-west Europe and north-east Asia, from Copenhagen and Hamburg to Tokyo and on round the world. The sleek, long aircraft, their powerful jet engines thrust forward like the four tongues of some monstrous dog, disgorge their hordes of tired, bored passengers, overfed and weary, lap up fuel by the ton, embark the passengers, revived by the clear air and the glorious sight of the mountains, and roar away again on their frantic schedules.

All this was far off in a distant and incredible future (like the oil finds below the waters of Cook Inlet) as the ships hove up from an Anchorage which was only anchorage to them, and stood down the Inlet to find some other way towards the north, regretting the loss of time. What they had to do now was to pass down the whole of the mountainous Alaska Peninsula and find a channel through the Aleutian Islands – a grim place at any time of the year, and rugged sailing. Here and there a puff of smoke indicated volcanic activity among the mountains: the much-tried *Resolution* took the ground lightly once but floated off without damage or trouble as the tide rose and lifted her. The natives brought out excellent salmon by the hundred-weight, which was most acceptable. They always asked for iron in exchange and were quite prepared to sell the sea-otter coats from their backs, apparently unaware of their value.

Sailing W.S.W. along the mountain chain – hundreds of miles of it – was trying, though not as bad as much of the Antarctic. The currents were strong and cold, fogs frequent, the mountains snow-capped, the land an iron-bound lee shore on which shipwreck was often imminent. Anchorages were few and far between, nor was there much time to look for them if the ships were to have any useful summer left for a drive to the north – if they could ever make it. At first Cook thought he was among islands and not embayed behind the arm of a continent, thrust out in front of him as if deliberately to hold him back. The almost constant poor visibility was exasperating. To anchor and wait for the weather to clear would be to waste the summer. Cook pushed on. Once, running with a fair wind through fog, making about five knots, suddenly the lookouts heard breakers ahead, under the lee bow. The always ready leadsman found twenty-five fathoms, then twenty-three.

'Down helm! Clew up everything! Haul the spanker to wind'ard! Leggo sprits'l and fores'l sheets, lively!'

She spun into wind.

'Let go anchor!'

So often, some variant of the same drill. It became almost routine. When it cleared, a horrified Cook (and Clerke, who was following close astern watching a fog-buoy – a piece of wood shaped like a scoop to throw up spray and give a mark to follow) looked astern to see, close behind them, two large pinnacles of rock with lesser rocks in between, among which both ships had by God's grace picked not just a passage, but the *only* passage, and that just wide enough for them to pass! He would not, said Cook, have tried to sail in such a place in broad daylight.

Acceptance of risk was inevitable. Racing tides, a profusion of rocks awash or just below the surface, scarcity of anchorages, absence of local knowledge among the few natives seen made sailing deep-draught square-riggers a nightmare which went on and on, and on. The island of Unalaska (Cook calls it Oonalashka) with its sheltered Dutch Harbour offered some refuge: through Uminak Pass nearby was free passage to the north, though hazardous. At Unalaska, later, Cook met a party of Russian fur traders led by one Ismailov, friendly fellows, marvellous cooks (they gave Cook a spiced salmon pie which reminded him pointedly how ill, by comparison, the English food was prepared: but it may have done his stomach little good) and willing to share such information as they had. M. Ismailov – Cook calls him Erasim Gregorioff Ismyloff – later showed the Englishman the Russian charts of the general area between Kamchatka and Alaska: but they had no knowledge of any North-West Passage, or tradition of such a thing. Their background was Asian and, beyond that, European. They were fur traders primarily. The English and Russian parties got along famously: Cook found the natives, too, courteous and polite wherever the Russians had been. They bowed on meeting or passing, and raised their fur caps.

Beyond Unalaska and the Aleutians were the Bering Sea and Bering Straits: beyond these, the sea called Chukchi, and then the impassable ice-jam of the Arctic. Cook probed on, noting the outflow of the great Yukon River, standing into Bristol Bay, Kuskokwim Bay, Norton Sound – all these in the far north, all of them indentations on the west coast of Alaska with always somewhere behind them the impressive wall of the vast, rolling mountains. Again the *Resolution* was leaking seriously: Cook forced the reluctant Whitby collier right up to the ice, enormous impregnable fields of it, not far from Point Barrow. If a sea passage reached east towards the Atlantic past here – as it did – he saw that it must be ice-jammed and useless, as indeed it was. He dodged the ships about between the rocky land and the great ice-fields, taking care not to be nipped. For the ice moves and the land is constant: a ship caught between would be ground to pieces. Cook extricated himself after a close look at the ice-field, the southern edge of which was some twelve feet high and the interior much higher – patient, impassive, endless, the last enemy in the North Pacific: but by no means silent. All along the edges it was noisy with the sea's surge: great floes ground together in the swell, lifting and smashing with a sort of malevolence which was frightening to watch and hear, as if the ice was grinding its massive teeth to get at the ships.

From a farthest north of nearly 71° – north of Siberia, close to the northernmost mainland point of North America, about the same latitude as Jan Mayen in the Greenland Sea –

Cook turned from the north Alaskan coast to beat patiently towards the Asian, to look for passage and examine the extent of the ice there. It was of no use: there was no way through nor hope of any, and it was getting late. A ship which allowed herself to be caught between the ice and the Siberian coast would be there for ever. Convincing himself that he had, for the time being at any rate, done all that any seaman could do, at last Cook turned back towards the Bering Strait on the Siberian side, charting the coasts both of America and Asia, and made slowly towards the good harbour at Unalaska. Walrus steaks (approved by Cook but at what cost we don't know: abominated by the mariners), excellent fish, masses of berries picked ashore and washed down with Cook's patent spruce 'beer' provided fresh food and kept the scurvy away. The walrus 'steaks' must have been tough. After being soaked for days, boiled for hours and then fried, they were still almost inedible even by hungry men.

It was early October before the two ships were back in Dutch Harbour, and the weather was atrocious. They had now been over two years at sea, often under appalling conditions. The *Resolution*, leaky again, was listed, the sheathing about the waterline removed temporarily and the seams caulked. The hold was cleared and restowed to give water that had leaked better access to the pump wells. Even worse than the badly caulked hull, in some ways, was the state of the rigging. The spars of these eighteenth-century ships were light and the cordage rigging was unsatisfactory, but the stuff put into Cook's ships for that voyage was scandalous. The trouble with naval refits in those days was that they were based on malpractice – poor materials used deliberately because contractors made profits that way which were denied them in any other (or smothered by the system of 'graft' – plain bribery – by means of which they were given their contracts in the first place). Poor materials bred poor workmanship. Cook found himself looking aloft ruefully, observing that the standing rigging, tackles, blocks and so forth dating back to her Whitby days which had survived the riggers stood up far better to rough usage than the 'new' stuff from the naval contractors. This was despite the fact that the merchant service material had been more than a year aboard when the ship was bought.

Typical of difficulties caused by plain bad cordage was the loss of almost a whole suit of sails when the bolt-ropes gave way in a squall. The strength of sails, especially in square-rigged ships, depends upon the roping at their edges – the bolt-ropes – and there is considerable skill in stitching these to the canvas. In the merchant service, it was the custom to use bolt-ropes of the best possible quality which would far outlast the canvas. When sails blew out, bolt-ropes stood: then you were left with at least the frame of a sail, and could sew and fit new canvas to that. (When I bought the ship *Joseph Conrad* in 1934 there were hempen bolt-ropes aboard dating back to the nineteenth century. In the bigger ships then they had long been made of steel wire.)

Cook must have looked aloft with disgust at the canvas flapping to pieces and the broken bolt-ropes writhing and thrashing at the men on the yards. If he swore a little – he probably did – it could be understood. It was a pity he could not send some of the scoundrelly contractors up to the yards, preferably naked, and let the torn bolt-ropes lash at them until they learned a lesson, which would take some time. But the ghastly contractors and those who fostered them were safe at home 12,000 miles away, doubtless earning knighthoods. Cook could reflect with satisfaction that at least the fore rigging and many of the vital brace and halliard blocks from the Whitby days were still giving good service. He had found them tossed ashore in a junk-heap, condemned – because no contractor would make money by leaving them in the ship – and had retrieved the lot himself. Mr Bligh – that excellent seaman who knew his rigging too – had formed a bos'n's party to carry the vital things aboard, and rig them when the contractors were not looking.

With much of the *Resolution*'s and *Discovery*'s rigging shattered, sails blown out, hulls leaky, and too many spars giving signs of trouble, it was necessary to choose a base where the crew could be refreshed and get on with a good refit. But where? At Petropavlovsk in Kamchatka, read his orders. This certainly was a convenient spot, being more or less adjacent to the western end of the Aleutian Islands: for the same reason it would make cold and grim winter quarters, with the ships snowed up and the mariners snowed up with them. The ships had to spend yet another season in the Arctic, and the Lord only knew when if ever they might reach England again after that. Besides, Captain Clerke, poor man, was suffering from tuberculosis, and Surgeon Anderson had recently died from that disease. So for once Cook read his orders (which were loose, anyway) with Byron's spectacles. No Petropavlovsk: he would make instead for those pleasant Sandwich Islands at whose Kauai and Niihau he had touched on the way north. They offered refreshment, sunshine and warmth, pleasant natives, an anchorage or two. His people needed sunshine.

'I had other reasons for not going to Petropaulowska,' * wrote Cook. 'The first ... was the great dislike I had to lying idle six or seven months which would be the consequence of wintering in any of these northern parts. No place was so conveniently within our reach, where we could expect to have our wants supplied, as the Sandwich Islands.' He had formed the opinion that these were an important discovery, and he could therefore make better use of the winter by examining and surveying them further.

Looking back with hindsight, it is a pity perhaps that he did not go to Nootka Sound and have a better look at Vancouver Island. Or find the Straits of Juan de Fuca, and sail in there to refit in some lovely bay on Puget Sound. The winter climate there is like New Zealand's, more fit for Europeans, perhaps, than tropical Hawaii's. Good mast and spar-making trees abound. Game, birds, fish, berries, spruce for 'beer'-brewing, the purest of fresh water all are plentiful. True, the local Indians were not Polynesians but, with Cook's approach, they would have been friendly enough.

But for some reason Cook scorned the idea that there was any Straits of Juan de Fuca, though he had been within a few miles of it when he named Cape Flattery.

He had obviously found in himself, long since, a great liking for the Polynesian islands and the Islanders beside whom the

* His spelling. The name means port of St Peter and St Paul.

west coast Indians had no appeal at all. But he might, one thinks now, at least have dropped in at Nootka Sound to cut some of those splendid Canadian spars: for the want of a good lower-mast was soon to kill him.

Away to the south the *Resolution* and the *Discovery* sped, pumps going, torn sails under repair, new bolt-ropes being laid laboriously from hempen strands made of oakum from her best cables (which being from the same contractors were equally useless), the officers and the mariners looking forward happily to yet another spell among the sunny islands.

Good harbours and tolerable anchorages are much scarcer than the landsman may think, even among the imagined serenity of the South Seas islands. The unclouded vault of heaven rests far too often over reef-littered, rockbound shores, which have stubbornly refused to open up arms to welcome the sea and provide good harbours. Open bays offer little shelter. Sandbanks, carried down by the hurrying waters as if to set up intentional barriers against the sea, bar too many river-mouths. Rocks, ledges, coral, and sea-sand give poor hold for biting anchors: onshore winds are ever more abundant, fresher, more persistently nocturnal: surf pounds the gentlest beach where the ship's boat runs in to land.

So it is also in the Sandwich or Hawaiian Islands. Indeed there is no such thing as a perfect natural harbour to be found in the lot, though some have been developed after immense expenditure – Pearl Harbor, Honolulu. For much of the year, the north-east trades blow home with strength all along the eastern and northern shores of the islands, funnelling between them like young and growing gales. The volcanic islands rise abruptly from the sea with no thought of ships' security, little allowance even for an anchorage. Honolulu on Oahu is perhaps the best port now, but the Honolulu of today is largely man-made and has not much in common with the indentation in the reef Cook knew as Honoruru. Even now, it is unsafe at times for smaller ships, for the sea can surge in when the *Kona* winds blow.

Going in there with a small square-rigged ship about the

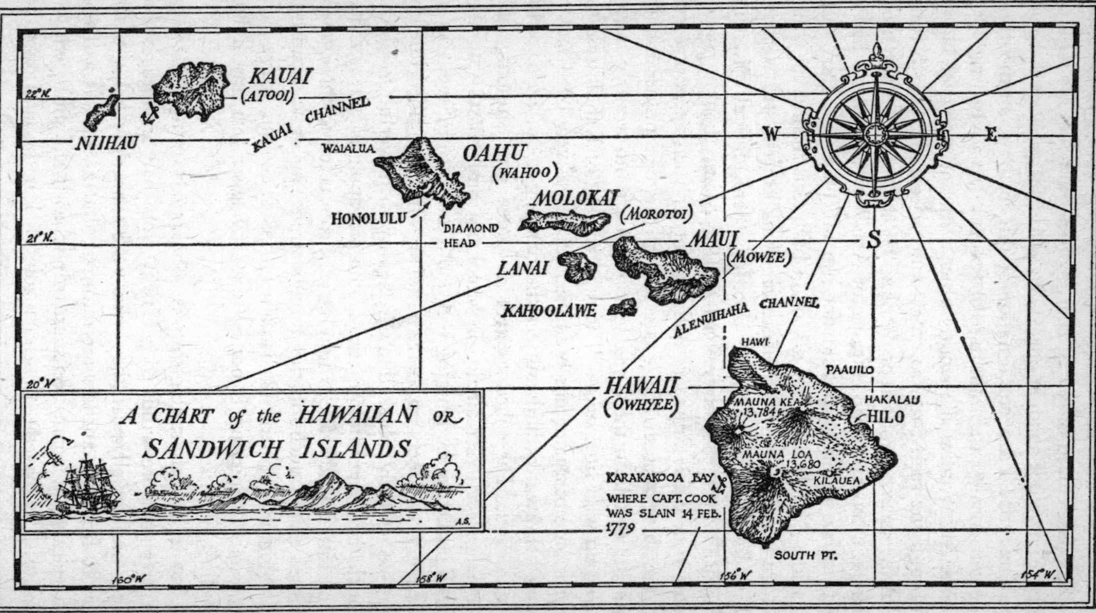
A CHART of the HAWAIIAN OR SANDWICH ISLANDS
KAUAI
(ATOOI)
NIIHAU
KAUAI CHANNEL
WAIALUA
OAHU
(WAHOO)
HONOLULU
DIAMOND HEAD
MOLOKAI
(MOROTOI)
LANAI
MAUI
(MOWEE)
KAHOOLAWE
ALENUIHAHA CHANNEL
HAWI
HAWAII
(OWHYEE)
PAAUILO
HAKALAU
HILO
MAUNA KEA
13,784
MAUNA LOA
13,680
KILAUEA
KARAKAKOOA BAY
WHERE CAPT. COOK
WAS SLAIN 14 FEB.
1779
SOUTH PT.
W
E
S
22° N.
21° N.
20° N
160° W
158° W
156° W
154° W.
A.S.

Resolution's size in June 1966, from San Pedro, the friendly U.S. Coast Guard officer in charge of the port met me with the advice to go out again at once as we could surge about there and be damaged: I saw confirmation of this in the giant moorings, big coir springs, the multitude of fenders. He advised me to go to the Ala Wai Canal, towards which a channel had been blasted through the reef. I did so, and found an excellent berth there. But the surf that rolls in on Waikiki beach sometimes also broke dangerously across the entrance of Ala Wai: at times I had to get out or, being out, could not run in. There are good yacht marinas: but a berth for a 400-ton barque is a different thing. There is quite a tide race at sea off Waikiki: the stiff winds in the Molokai channel make beating there difficult. Lahaina Roads, near the south-western end of Maui, though popular later with whale-ships, are nothing but an open anchorage providing little real shelter – safe enough in the summer months but impossible in the *Kona* season when south-westerly storms blow up with little notice. The modern port of Kahului, on the north side of Maui, has been created expensively by means of long breakwaters built out on the fringing coral. So has Hilo in Hawaii. These breakwaters at times give shipping only partial protection. Kamalo, on Molokai's south coast, is nothing but a pocket in the fringing reef and a pier for pineapple barges (heavy steel vessels up to 12,000 tons). These are filled quickly, and as quickly plucked away again by powerful tow-boats. Much the same thing applies at the pineapple island of Lanai. The inter-island trade is now largely conducted in these towed barges, and much of the foreign trade in ugly steamships carrying huge loads of containers. A 12,000-ton barge or a 15,000-ton steamship is not bothered by lesser seas and lighter swells.

But a 400-ton sailing-ship is. She is also vulnerable: she prefers off-shore winds to get away from bad anchorages. She likes most of all a safe bay where the chances of being blown ashore from her anchors are minimal, and a ship has a chance of beating out against on-shore winds, as if she must, without becoming embayed – that is, unable to clear the land on either tack.

And so Captain Cook sailed down from the Arctic, and beat the ships for many days round the chain of islands. From Maui to Hawaii, to Oahu, Molokai, and back against the wind again to the big island which he called Owhyee, he found no good harbour. He kept the two ships at sea continuously for eight weeks, heaving-to and trading for fruit and food with canoes which came happily out, but never finding good anchorage. Near Hilo Bay more of the infernal bolt-ropes parted, leaving the *Resolution* without sufficient canvas off a precipitous lee shore. At last, on the western side of the big island – Hawaii is much the largest island in the group – he noted a shallow bay. Mr Bligh, sent to sound and investigate, reported it as possible, a bay two miles wide and a mile deep, wide open to the south-west storms (already Cook knew about these) but otherwise easy to sail into and simple to leave. The holding ground seemed good.

Cook sailed in, Clerke with the *Discovery* following. Their welcome was extraordinary: canoes by the hundred splashed about as the ships moved slowly towards anchorage. It seemed that most of the island's people came out – thousands of them. Those not in the canoes swam in the clear water like great shoals of golden-skinned fish, splashing away, reaching for lines thrown to them, clambering happily aboard. The *Resolution* anchored about quarter of a mile off the north-eastern side of the bay and the beach here, too, swarmed with people. For the time being, despite the extraordinary warmth of tumultuous welcome, the sailors attended to the proper securing of the ships, mooring, unbending the sails, striking the lighter yards, the topgallant masts and topmasts, to give the ship a thorough overhaul. The warmth of the welcome almost made the seamen glad they had found no icy North-West or North-East Passage across the top of the world to probe towards England. The scene there at Hawaii was colourful, warm, welcoming and brilliant, with the brooding mountain-mass of volcanic Mauna Loa dominating the lovely island and the warm sea lapping gently along the beach – a bewitching sight for sea-weary eyes.

It was 17 January 1779. The Hawaiians said their bay was called Kealakakua, pronouncing each syllable distinctly as is the

Polynesian way. Cook heard it as Karakakooa. All these islands from Kauai to Hawaii were new discoveries. Cook was delighted with them: so were the seamen. There was further persistent thievery: but hogs, greens, roots, coconuts and fruits were abundant and not only fairly traded but brought as gifts. This was new, and acceptable. So many hogs were offered that Cook salted much of the pork in casks for use in the summer. His eyes were set even from this pleasant tropic island on, far beyond, the ice-bound Arctic. Sails were patched and freshly roped, spars and masts repaired or replaced, cordage refurbished. The caulking hammers sang all day as decks and sides were dealt with by the carpenters' gang. Boats plied busily bringing off fresh water, vegetables, firewood. An observatory was set up ashore.

Whenever he landed, again the Polynesians prostrated themselves before Captain Cook, as they had at Kauai the year before. Now in addition there were mystifying ceremonies, processions, long harangues which sometimes seemed to take the form of services, and formal exchanges of apparently valuable gifts. Cook gathered that he was Orono or, more likely, *an* Orono,* for he thought this to be a kind of Hawaiian title; it was all part of very friendly and useful relations and otherwise meant little or nothing to him. The supreme chief (virtually king) of the island, one Kaleiopu'u or Kalaniopu'u, was rowed round the ships with solemn ceremony accompanied by some of the high priests and other chiefs, and visited aboard. Cook was presented with some of the chiefs' magnificent red-feather cloaks and extraordinary Grecian helmets, obviously most expensive and reserved for the highest chiefs. When Cook landed, senior priests unobtrusively attached themselves to his party. They went with him everywhere. Part of the reason was to make sure that he was treated with proper respect. Cook did not suspect that, having built him up, they were determined that the common people must accept him at their valuation.

*In 1788, Andrew Kippis – an early and reliable biographer of Cook – was describing 'Orono' as 'a title of high honour', bestowed on Cook. It was a long time before the truth was established.

Exactly what was an *Orono*? Or, perhaps, who was this Orono?

'Some of these ceremonies,' said Lieutenant King ('Tinnee' to the Polynesians, with whom he had an unusual facility for making friendly and far from shallow contact) – 'Some of these ceremonies seem to border on adoration.'

They did: for the 'Rono' was in fact Lono the god, a cheerful, earthy Hawaiian of long ago who had been exiled, and prophesied his return in a large island with trees, bringing gifts, including swine and dogs. Well, here were the 'islands' complete with 'trees', and swine and dogs. Here, too, was the tall, imperious but friendly reincarnation of Lono, in the shape of Captain Cook. What else could he be? The very day before his first arrival with his 'islands' off Mauai, the king of Hawaii, Kalaniopu'u himself, had been victorious in a battle there – obviously because the great Lono was in the offing, come to his support. Honaunau, on Kealakakua Bay, is the ancient walled city of refuge, a much revered place in old Hawaii. Time and setting were right for the return of a 'god'. The sailors in his 'islands', too, were no ordinary men. The astonished Hawaiians noted them carrying fires burning in their mouths (pipes): when they needed anything they reached into their extraordinary skins (jackets) and took it out: some had heads horned like the moon (officers' uniform felt hats of the period): they could take off the tops of their heads (wigs) and mop their faces with cloth of an impossible softness (lawn and linen handkerchiefs). Whence could they have come but the abode of the gods?

Maybe. They took an awful lot of feeding and had other distressingly human traits. The priests and the chiefs had to make constant levies to keep up with them. In time this sort of thing could become irksome: but the amount of food and gifts brought to the ships was none of Cook's asking. He had no idea it was mainly priestly imposts, taken from the people for the god Lono's happiness – and, perhaps, propitiation, to say nothing of the maintenance of despotic government in the hands of the priesthood.

The priests, the chiefs, and everyone else heaved great sighs of relief when at last, early in February, with their sails and rigging repaired as well as they could be and the ships better stored than had ever been in the Pacific before, the *Resolution* and the *Discovery* took up their anchors, Lono spread his banners high on his trees, and moved out of the bay. It had been a great visit, but the local larders, gardens, sweet-potato patches and pig-sties were considerably depleted.

He was very pleased with Kealakakua, Cook wrote in his journal, and delighted 'to enrich our voyage with a discovery which, though the last, seemed in many respects to be the most important that had hitherto been made by Europeans throughout the extent of the Pacific Ocean'.

The last? What did he mean by that? He had then another whole year of further voyaging before him. Had he some premonition? Odd words, indeed: the last, certainly – and the last words, too, which Cook was to write in his journal.

It was the *Kona* season, the time of storms. The *Resolution* had not sailed far before a hard blow caught up with her, a close-reefed tops'ls business, with the wind screaming in the rigging – a bit of a blow but nothing more than she should have stood easily. The following morning showed an ominous working of the fore topmast and topgallantmast. As the ship slid down a trough or rose on a crest and rolled, the topmast and topgallantmast rolled sickeningly further. With a jerking motion, they rolled more than the ship did.

This could mean one thing only – serious rigging damage, for the tenth time that voyage. Oh wartime contractors, hurry-up riggers, purveyors of imperfect masts and sub-standard cordage, a curse upon you all, diabolical handicappers of ships, murderers of men! Cook looked aloft, grimly. It was obvious where the seat of the trouble lay. The fore lower-mast was gone again, sprung – split – at the head. Some eyes of the shrouds had slipped, where a grommett had parted on top of the hounds, leaving the mast improperly supported. Then the topmast, stepped above, began to move, to 'work', increasing the damage. That foremast was a comparatively new spar from Nootka

Sound, but strained from the Arctic voyage. It had been repaired and carefully fished – lashed with splints – in Kealakakua Bay. Inspection aloft showed serious damage which could not be repaired at sea. The ship must find harbour again. But where now? Cook was against a return to Kealakakua. His purpose was to make a survey of these new Sandwich Islands. There was still good time for that before the summer return to the Arctic, but not if weeks were to be lost beating about the islands looking for a harbour.

With the lower-mast sprung, the foremast was next to useless and could carry no sail above the course. Down came the topmast and topgallantmast. Down came the main topgallantmast too. The ship could now set only six or seven sails instead of her normal twelve. Handicapped as she was, Cook continued to sail, fighting wind and sea, desperate to find anchorage. The whole fore lower-mast would have to come out, and that could only be done on an even keel, in harbour.

There was no anchorage other than Kealakakua. Reluctantly, Cook returned.

But the white banners of Lono on his floating islands stirred no welcoming chords this time. A sullenness came over the golden-skinned Hawaiians, for they feared further imposts for the sustenance of Lono's men from their despotic priests and chiefs. Neither they nor any other islanders were accustomed to produce a surplus beyond their needs, for possible trade. Where nature was so bountiful, sufficient unto the day was the general rule. Lono's first visit had been a strain.

Now he was lifting out his 'trees'. Cook wasted no time unstepping the damaged mast and floating it ashore for thorough repair. How long would he and his 200 followers stay? The priests, resigned to do their best, offered some pale welcome: the citizens threw stones: the chiefs took counsel. Cook told them as well as he could (missing the dead Anderson who spoke the best Polynesian) why he had returned and what he had to do. They seemed to understand. But there were 'incidents'. Thieving became bold, persistent, and serious. There was some retaliation.

There was increasing awareness that whatever 'Lono' Cook

might be, these seamen were no gods. They were mortal: they could die. An old seaman named William Watman, a former retainer of Cook's at Greenwich who insisted on following his captain back to sea, had died not long before the *Resolution* sailed. Cook buried him ashore, the natives watching silently with some shock. On such occasions previously bodies had been quietly committed to the deep off-shore: perhaps the old man had requested a land burial. It was allowable that the lesser gods sailing with Lono should be boisterous, carnal, act as the mortals in whose guise they were appearing: but it was odd for them to die. It was odd, too, that they had brought no women: where were Lono's women? He used to like women when he had been in Hawaii long before.

Thieving became an ever greater menace. The natives developed a means of prising out the long nails which held the sheathing to the ship's bottom by using a flint fixed to a stick. Armed with this, they dived under the ship and dug out vital fastenings as fast they could and for as long as they could hold their breath. This was intolerable. It was little use to repair ships which were being pulled to pieces. One who defied warnings was flogged as an example. It made little difference. The standard of vigilance to prevent all theft was impossible. The majority of the crew had to be ashore, working on the mast, which was found to be in worse state than feared. The heel was rotten, too. Cook's temper at times was ragged.

Day after day, incidents increased. A pursued thief was not caught. An affray followed which did not go well, for Cook – never relying on unearthly attributes which he neither believed nor understood – would not use the superiority of the ship's firepower. He was not come to murder. One night a large boat, the *Discovery*'s cutter, was stolen from her moorings close by the ship.

This was more than Cook could stand. The cutter was vital and could not be replaced. He had a regular drill for serious thefts of this kind – to seize the highest chief available and hold him as hostage until the article was returned. It had always worked. It was bloodless but salutary – the more so as it was often a chief who had instigated the robbery, and chiefs were usually related. Seize one: the guilty would act.

As a secondary precaution, armed boats were sent to prevent canoes from leaving the bay until the cutter was returned.

Cook, in full uniform and carrying his personal double-barrelled shot-gun, was solemnly rowed ashore. With him was an armed guard of nine marines under Lieutenant Phillips, their officer. Lieutenant King went in with another boat to give support. King Kalaniopu'u was ashore and was to be the hostage. Cook looked worried. There might be trouble.

The nice old king, told of the theft (with which it was obvious that he had no connexion), agreed to come with Cook to the *Resolution*, without question. Up he got at once and began to walk with Cook to the beach, very calmly, two of his young sons accompanying him. They reached within twenty-five yards of the boat. The boys ran ahead and jumped in.

Meanwhile a large and excited crowd was assembling, alarmed at the sound of musketry in the bay where some canoes had been stopped.

A woman got between the beach and the King. She was his favourite wife. With tears and loud sobs growing steadily louder, she begged him not to go any further. Several young chiefs joined the group menacingly. The crowd closed in. Lieutenant Phillips noted some of them picking up stones. Others darted into houses, coming out again with spears and clubs, and fastening their war mats – a kind of breast and back-plate woven from coconut fibre, proof against spears and stones.

The King began to continue with Cook. The wife wailed. Two young chiefs gently pushed the king down to a sitting position in order that he could not walk. Cook urged him on: the chiefs held him.

Lieutenant Phillips quietly drew up his marines in line along the beach, at the ready.

Suddenly there was a great hubbub from the crowd, a roar of anger. An important chief, they shouted, had been killed in the gunfire on the bay.

Cook left the king, telling Phillips that they must go back

without him to prevent serious bloodshed. A warrior in breast-plate rushed up to the Captain, menacing him with a stone in one hand, dagger in the other.

'Put those things down!' ordered the captain.

The warrior made ready to fling the stone. Cook fired at once. One barrel of his gun was loaded with small shot. He used this. The pellets bounced off the tough mat. The warrior laughed and yelled defiance, and came on with his dagger. This time Cook fired the other barrel, loaded with ball. A warrior dropped.

A general attack with stones began at once. The marines fired: but the warriors having noticed the captain's first shot fail to do harm, and not waiting to see the effect of the second, pressed on. Seeing the muskets flash and now thinking the brief flash of flame was all the guns could manage, they ran to the water's edge, wetting their mats, shouting defiance. Others rushed the marines before they had time to reload, struck down four of them and wounded others. Now the seamen in the boats came to their support, opening up.

For a moment Cook stood there, facing the shouting, suddenly blood-hungry crowd: while he faced them none struck at him. He did not reload.

He turned to make for the boats, raising a hand to command a cease-fire. What he said could not be heard though he had reached the water's edge.

As he turned away a warrior rushed at him from behind, clubbing him violently. He sank to his knees, half in the water. The warrior stabbed, and stabbed again. A ghastly roar broke out from the maddened mob as Cook went down. Men rushed into the sea, stabbing, clubbing, holding him under the water. Once he raised that noble head, and looked at them – once only. Then he dropped back insensible. They dragged his body ashore and stabbed and stabbed in frenzy, seizing the daggers from one another's hands as if each must assure himself that the blood flowed.

This desperate business took seconds only – a ghastly, irretrievable flare-up, unpremeditated, utterly unnecessary.

No one, for the moment, could now stop the sailors and the

surviving marines. The warriors learned that their breastplates were not armour.

The beach cleared. The boats pulled back with their incredible news. When it was heard, a stifled silence, a numbed and hopeless bewilderment, a great sorrow descended on the ships and over Kealakakua Bay.

CHAPTER FOURTEEN

The Enigma of Captain Cook

IT was tragic that James Cook should die in a hot-headed fracas, killed by Polynesians with whom elsewhere he had done so much to establish good relations. He had had too little time to become known to the Hawaiians, and the unnecessary business of confusion with the ancient Lono was no help. As for that, it is obvious that Cook had little if any idea what it was all about. Long research among the Hawaiians themselves, years afterwards (though while witnesses were still living) when the Hawaiian language was thoroughly understood, was necessary to establish the truth from the Hawaiian point of view. What chance was there for Cook to realize that some of the earlier ceremonies were perhaps meant as a form of Hawaiian idolatry? He had been received as a great white chief before: he *was* a great white leader, the representative of his King, the leader of a great expedition. As for the priestly mumbo-jumbo, he suffered it as part of the Polynesian way of life, for priests and chiefs seemed to be accepted as descended from the gods. So he put up with it as necessary for good relations, and for the welfare of his crews. He had done that before, too. It was the only sensible policy. Since, as far as any of the Englishmen could understand, all the chiefs claimed descent from the gods and based their right to rule on this, such a relationship carried little of the godlike about it. The grant of similar status to the white leader was only proper.

Moreover, the good relations which Cook strove always to foster with natives everywhere were essentially between human beings as such – the European and the Polynesian, Maori, Melanesian, Australian Aborigine, Eskimo, Canadian Indian, Patagonian – on a human basis, never on the false and stupid basis of the acceptance of any form of the supernatural, the godlike, or belief in such notions fostered for any purpose.

It was unfortunate that the return to Kealakakua had to be made, and far more so that Cook was by then a man under strain, at the end of too many, too long voyages. A man does not drive himself, unrelieved over years, in command of ocean-wandering square-rigged ships without paying a heavy price for it in mind and body, no matter how strong his constitution is thought to be.

Now Cook was dead: but the voyage must go on as he had planned it. The horrified seamen, once the tragic clamour died down, got on with the urgent work of repairing the foremast and the sails. Almost the entire suits of working sails from both ships were ashore where they could be spread out and mended more easily than on the restricted decks. Now they were brought back aboard. So was the mast, a bulky, solid piece of timber almost as long as the ship. Working as quickly as they could, anxious to be gone from the fatal bay, the carpenters' gang spliced in a new heel, fitted new cheeks cut from a spare anchor stock, repaired the mast-head, and then hoisted the mast and restepped it. To do this they had to rig heavy tackles from a pair of sheers, a sort of jib-less crane contrived by lashing two spare topmasts together, their feet astride the decks and the lifting tackle rigged where their heads crossed and were lashed. Then the mast was lifted, nicely balanced, and the heel threaded into the holes for it in the main and 'tween-decks, and lowered to the kelson on which it stood.

All this was slow, laborious work, made much more difficult and dangerous by the rottenness of the cordage. The hawser rove off, for the hoisting tackle parted six times while the men were heaving at the capstan. They repaired it each time, and persisted. The mast was stepped and rigged, and the top-mast set up above it. There was no timber accessible on Hawaii to provide new masts: the patched-up old had to do.

Captain Clerke, from the *Discovery*, took Cook's command, and he appointed Lieutenant Gore, first lieutenant of the *Resolution*, to his former ship. The lieutenants stepped up a rank as necessary, and a midshipman was promoted to bring them to complement. All attempts to get Cook's body and those of the four marines who had died with him failed, though a few of the captain's bones, his hands, and a piece of charred flesh were brought out by a sympathetic chief. Then the bay was cleared, all work stopped, the men fallen in aboard, dressed in their tattered best, officers and marines in full regalia and the remains solemnly committed to the deep with quietly moving ceremony and a salute of ten guns. Captain Clerke read the service from Cook's Bible, so often used by him. (The ships, since they were rated in the Navy as sloops, carried no chaplain. None was allowed for such small ships.)

Afterwards, with the ships rigged, the sails bent, sufficient fresh water and fresh food aboard (for the people on the side of the bay opposite that where the affray had occurred, after a while, quietly resumed trade), Captains Clerke and Gore picked up their moorings and sailed. It was a grey morning, gusty with rain – a day fitting the seamen's mood and, indeed, the Hawaiians: for they also, when their tempers cooled, lamented the death of Captain Cook.

No trace was found of the stolen cutter. Long afterwards, it was learned that one of the lesser chiefs had stolen it and had it broken up for the sake of its metal fastenings.

Captain Clerke was dying from tuberculosis, contracted before sailing from England in a debtor's prison to which usurers had committed him for alleged responsibility for the debts of a relative, but he pursued the voyage just as Captain Cook would have done. The ill-treated *Resolution* soon leaked again, at times of bad weather up to a foot an hour even with the pumps going. But Clerke pressed on by way of Kamchatka to the Arctic for one more endeavour there. It was useless. The ice-fields of 1779 were larger and further south than in the previous year. There was no hope of breaking through – no hope of getting anywhere by that route. The ships' state, bad when they left Hawaii, steadily deteriorated. The seamen had to nurse the rigging, suffer wet in the quarters, care for the sails as if they were made of linen and the running rigging as if it were cotton thread. One day a lot of timber floated past – it was part of the *Resolution*'s own sheathing. The cold was intense. The *Discovery* suffered hull damage in the ice and was in even worse condition. Not matter what else happened, Cook's anti-scurvy drill was rigorously continued – airing the ships with charcoal fires, washing out the quarters with a vinegar-and-gunpowder solution, serving out large portions of sour-krout, portable soups, etc. to all hands, insisting on high standards of personal cleanliness.

In the midst of all this, having sailed the ships as far as it was possible for sailing-ships to go, Captain Clerke died. He had been in the Navy since he was a child, had seen action on the American station and been flung as a junior mizzen topman aged twelve out of the top when the mizzen-mast was shot from under him, was a midshipman under Byron in the *Dolphin* and Master's Mate (late lieutenant) with Cook in the *Endeavour*. The worst possible place that he could have gone in his state of health was on an Arctic voyage: he would have been justified if he had sailed back to England from Kealakakua Bay. He might have lived, then. But he was a sailor in the Cook tradition. He was thirty-eight years old. Lieutenant Gore now took command of the *Resolution* and Lieutenant King of the *Discovery*.

Poor Clerke had asked to be buried ashore, and not in those

icy seas: Gore therefore carried the body to Petropavlovsk. Here the ships were again repaired as thoroughly as possible, and fresh provisions including bear meat, large quantities of salmon and sixteen bullocks were embarked. Having now done all that such ships could possible manage, the *Resolution* and *Discovery* sailed for home. Among many other important shortages, they had long been without a current nautical almanac, and had to work out their sights laboriously, the more so as the copy of Harrison's chronometer, after more than three years of solid work, stopped and could not be got going regularly again.

It was early in October 1779, when the ships got under way homeward-bound from Petropavlovsk, firing a salute of twenty-one guns in honour of the Empress of Russia (and for the friendly local Russians) as they made sail. One thing Gore did there was to send a letter overland to Admiralty in London with copies of Cook's report of proceedings and his own dispatches on Cook's death. Off this went by dog-hauled post-sled. Across the whole of Siberia and all Russia team after team of good dogs followed each other, hauling the mail-bag over snow and ice. Later horsemen took up the task when the sled could no longer be used, then coastal shipping across the North Sea to England.

Six months later, the dispatches unexpectedly arrived in London, to the consternation and dismay of all Cook's friends. Another six months after that, in early October 1780, the *Resolution* and her consort arrived back at last in England after a voyage of four years and three months. On the last leg, they had staggered down the whole western side of the North Pacific and Indian Oceans, taking the China and Java Seas in their stride, and sailed again the whole length of the Atlantic. In the end, not at all sure of their reception by possible French privateers in the English Channel and hearing from a fisherman off Galway Bay in Ireland of the marauding John Paul Jones in English waters the previous year, Gore took no chances. Newspapers from England and America found during a call at the Portuguese port of Macão, in China, had told him of the French safe-conduct orders for the ships: but these might not be binding on privateers. As for Jones, the same papers spoke of his exploits and did not mention that Captain Jones was a friend of Benjamin Franklin's. In any event, after taking the frigate *Serapis* (but missing her convoy) with the sinking wreck of his *Bon Homme Richard*, Jones was gone.

Gore sailed the ships right round Ireland and, with a call at Stromness in the Orkneys, beat down the North Sea to the Thames. The Whitby ships sailed for the last few hundred miles on their old collier run, from the Yorkshire ports southwards. This time the coasting ships recognized them and cheered them as they wandered past with their battered masts, their bleached, sea-weary hulls and their threadbare sails, and the clear water pumped from their leaking hulls streaming from their decks and down their sides.

On the way down the North Sea, two elderly men of the *Resolution*'s crew died, perhaps worn out. But not a life was lost to scurvy in either ship over that whole, incredibly long voyage – the *third* time James Cook and, at last, his shade had performed that eighteenth-century miracle. It was the greater miracle as his dietary methods, though helpful, were not really rich in anti-scorbutics, apart from the use of fresh food and fruit. It was the cleanliness, the insistence on clean living, the new idea of fresh air circulating in formerly fetid quarters and, above all, the constant drive to maintain these standards and the idea that scurvy *could* be defeated – these were the reasons for Cook's success. Anyone with the necessary strength of character and qualities of leadership could apply them: even after Cook showed the way, few did.

There was another factor that helped Cook. Though he made longer voyages than had ever before been made, he took care to break them whenever he could, by calls at islands. When he called anywhere, his first quest was fresh food even if it were only grass to mix with portable soup, sour krout and such stuffs to force down the seamen's throats. He did not keep his ships unnecessarily long at sea, though the landlubber Forsters might have thought so and found – or made – some publicity for their foolish lamentations. Another of their criticisms was that the Captain would not divulge his intended course to them, and would not tell them where he was going. This was in the white labyrinth of Antarctica. How could he disclose a 'course' he did not know? Did he or anyone else know where he *could* go? It was because nobody knew that he was there, to sail on and to stay until the mists rolled back or formed themselves into ice impenetrable.

In the end, the ships came back to England very quietly. Their principal news was already known. Their officers and men dispersed to war duties, for the War of the American Revolution was going badly and still had three years to run. The ships themselves were taken into dock and refitted thoroughly – at last! No criminal contractor, no purveyor of sub-standard cordage or skimper of shipwrights' work was hanged or shot, for such wretches were exceedingly prosperous and politically powerful. Some indeed, were in the government. The miseries and dangers they caused to seamen went unavenged. Those who had been most deeply wronged were dead. The others thanked God to be alive, and were aware that they were without political influence.

The *Resolution* was converted to an armed transport, sent to the East Indies, and disappeared from the records. An Australian report a hundred years afterwards that she had finished her days as a coal hulk in Rio de Janerio seems unlikely and cannot be confirmed. The smaller *Discovery* was reduced later to a reception-hulk for convicts awaiting transportation to Botany Bay. Dismasted, her bluff old hull all tarred, her masts taken out of her and her decks over-built with ugly houses, she sat on the mud at Deptford for years until increasing disintegration caused her to be broken up. As for the other ships of Cook's noble, bluff-bowed quartet, Furneaux's *Adventure* went back to Whitby owners from whom she sailed for thirty-five years. At last, in 1811, she found herself in the St Lawrence, that scene of Cook's first triumph. This time the river determined to keep her there and rose up and wrecked her. The *Endeavour* alone still has some existence. She was wrecked or damaged by stranding and afterwards allowed to fall to pieces at Newport, Rhode Island, and some of her timbers are still there. She was the French whale-ship *Liberté* – really Franco-American, for apparently the French provided a working subsidy and the New

Englanders the whaling skill – when she arrived there some time in 1795. She should have made a good whaler, but she was either seriously damaged or, in time, condemned. Part of her original sternpost is preserved in the museum of the Newport Historical Society, who provided a piece to be built into the proposed *Endeavour* replica memorial to Cook in 1966.

There was what seems now a curious reluctance to do adequate honour to the memory of Cook, or perhaps a failure among his contemporaries to appreciate just how great a man he was. Mrs Cook was voted an adequate pension – £200 a year, with £25 a year in addition for each of the three children then living – and a half-share in the profits on books of her husband's voyages, based on his journals and officially sponsored. A great many books were not (and one or two hasty works were downright misleading), but Mrs Cook became a wealthy woman, for those days, on her share. Her needs were small, poor woman: soon she had only herself to keep. Cook's family life was brief and much interrupted: Mrs Cook's was brief and tragic. Cook had been at home very little with his wife. He was allowed no shore appointments (apart from those few months at Greenwich Hospital) and wanted none. His periods of what should have been leave were taken up with works on the charts and surveys of previous voyages and arduous preparations for the next.

There were six children: only one survived to manhood. None lived a full life. Two boys died in infancy, and a girl at the age of four. Two boys, James and Nathaniel, survived long enough to take up naval careers. Nathaniel went down with all hands aboard the frigate *Thunderer* in a West Indies hurricane in September 1780, not long after the assassination of his father. He was a midshipman, aged sixteen. His brother Hugh, scholar of Christ's, Cambridge, died of a fever there, aged seventeen. A few months later James Jr, last in the line and then a commander in the R.N., was drowned out of the *Spitfire* sloop, to the command of which he had been appointed. He left no children. With him, the James Cook line died without having had much chance to become established: a grant of arms to the family made by the King was quite wasted, though such a grant was then an honour not given idly. Mrs Cook, overwhelmed by all these deaths one after the other, retired to Clapham, in London. There she lived with her memories into the steam age and died in 1835, at the age of ninety-three. Her cousin Isaac Smith – the same who as a midshipman was first to leap ashore at Botany Bay – lived with her, a superannuated kindly old admiral: but he predeceased her, too. There is at least a second-hand account of the old lady, written by Sir Walter Besant (1836–1901) in a biography of her husband,* at a time when a few old friends of Mrs Cook were still alive. Among these, a Canon Bennett, of Maddington Vicarage, Devizes, was most informative.

'She kept her faculties to the end,' the Canon said. 'I remember her as a handsome and venerable lady, her white hair rolled back in ancient fashion, always dressed in black satin, with an oval face, an aquiline nose, and a good mouth. She wore a ring with her husband's hair in it: and she entertained the highest respect for his memory, measuring everything by his standards of honour and morality. Her keenest expression of disapprobation was that "Mr Cook would never have done so".'

Like many sailing-ship captains' wives, she could not sleep on nights of high winds, thinking of the men at sea. She kept the anniversaries of the deaths of her husband and three boys as fast days in meditation and reading from her husband's Bible. Her home was 'good, filled with old furniture of the type called Louis Quinze; it was also crowded and crammed in every room with relics, curiosities, drawings, maps, and collections brought home from the voyages. ... On Thursdays she always entertained her friends to dinner, which was served at three o'clock.'

One thing the house did not contain was anything of Cook's letters or private papers of any kind, for the old lady most carefully destroyed the lot. They were private, not for prying eyes: she had obviously little sense of history or of the real stature of 'Mr Cook', and a very real sense of duty towards him. One wonders how well she really knew him: how well, indeed, anyone knew him.

* *Captain Cook*, by Walter Besant. MacMillan & Co., London and New York, 1890.

Mrs Cook left directions for her body to be taken to Cambridge for burial in the centre aisle of the church called Great St Andrews, not far from Christ's College. Here the remains of her scholar and her sailor sons, Hugh and James, had long been waiting for her. She left money to erect and maintain a memorial in the church to her husband and their family. There today the tragic record stands, in memory of Captain James Cook, 'of the Royal Navy, one of the most celebrated Navigators that this or former Ages can boast of'; of Mr Nathaniel Cook, 'lost with the *Thunderer* Man of War, Captain Boyle Walsingham, in a most dreadful hurricane'; of Mr Hugh Cook, 'of Christ's College, Cambridge'; of James Cook, Esq., 'Commander in the Royal Navy, who lost his life on the 25th of January, 1794, in going from Pool to the *Spitfire* sloop of war, which he commanded': the infants Elizabeth, four years; Joseph, one month; George, four months. Finally, Elizabeth the wife and mother, who survived her husband fifty-six years. Part of the Cook arms is shown beneath. In the centre aisle a slab above their grave names the three Cooks who lie there.

As far as one knows, this is the only memorial to James Cook in a church anywhere. There is none in Westminster Abbey. A move to have a suitable memorial put up there came to nothing in the eighteenth century and has not been revived. His old friend and admirer, Sir Hugh Palliser, erected one of his own on his estate of the Vache, Chalfont St Giles, not very far from London. It is a large stone, or tablet, with a long inscription to 'the ablest and most renowned navigator this or any country hath produced'. The memorial stands today on National Coal Board property. There was a move recently to transfer it to Australia.

The Copley Gold Medal awarded by the Royal Society to Cook before his last voyage for his paper on his measures for seamen's health, reached Mrs Cook by messenger. So did another gold medal, especially struck by the Society to honour the great seaman's many achievements, with a letter from its President, Sir Joseph Banks. 'His name will live ever,' wrote Sir Joseph. Copies of the gold medal were also sent to the King, Ben Franklin, Lord Sandwich, and Sir Hugh Palliser.

There was no award of any part of the £20,000 North-West Passage prize money for Cook's estate nor for his officers and seamen, though they had in fact made the final contribution to this enigma. After Cook, it was unnecessary to send another ship in search of the imagined Passage before the days of powerful ice-breakers and small vessels especially designed to be gripped in the drifting ice and stay with it. Ships were sent from the Atlantic side and lost, their men with them: none again went to the Pacific. Obviously Cook scarcely thought of the prize: he never mentions it. Nor do Clerke, Gore or King: but the sailors might have had a bonus. The unfortunate truth was that they and their ships slipped unobtrusively back to England in the midst of an expensive and unnecessary war not going at all well, and soon to be lost. Cook was dead. So was Clerke. Dead officers are dead officers. One again, the great seaman's achievements were for posterity, not his harassed contemporaries who could not and did not appreciate them. To them, they all seemed negative. There was no ready-made exploitable great land waiting in the South Pacific – nothing but ice, island-groups and islands, with the two islands of New Zealand and the considerable land of eastern New Holland thrown in for what they might be worth. In the contemporary view, this was not much. These discoveries were not brought back to a crowded England full of would-be colonists and emigrants. There was still room of a sort for all. Pilgrim Fathers, Puritans, and many Quakers had found new homes in America which was in process of being lost after a considerable investment in it. In these circumstances, the idea of expensively colonizing a distant New Holland and New Zealand was taken seriously by few indeed, at least for the time being. The cost of the American war was sufficient to carry.

There were denigrators of Cook, of course, even calumniators. It might seem difficult to vilify the memory of the humane Cook, but there was at least one so-called missionary in the Sandwich Islands who set out on a deliberate campaign to falsify the record and defame his memory. Having read, perhaps, one of the works of the unstable corporal of marines, Ledyard,* in which Cook's murder at Hawaii and the events leading to that tragedy were greatly misrepresented, he accused Cook of being a willing participator in idolatry, lecher, a contaminating influence and bringer of disease. Cook was the man who played god and the Lord's wrath descended upon him – good stuff to bring home a fearful sense of the Divine retribution to the happy Hawaiians, and to denigrate the great English seaman. As for the 'lechery', this was a particularly sly and baseless charge. In a loose and free-loving age, Cook was noted for his unique chastity. In the islands he named Friendly, as part of his welcome once a bevy of the local maidens was presented before him to make a choice of a temporary harim, as was the custom. They were shapely, luscious, young, willing. He smiled, and rejected the lot, to the astonishment and noisy chagrin of the older ladies who had assembled them. What manner of man was this? For he was beyond them – beyond a type of missionary, too.

The reverend gentleman, not satisfied with the harm his narrow kind had already achieved in the Islands and the sound foundation some of them had laid for more, carried on a vicious campaign of calumny. He was in a position to poison young Hawaiian minds, for he also professed to be a teacher. The gentleman may be left to the infinite mercy of the God he served so oddly. There was an element of anti-British propaganda behind his distorted diatribes, for the commercial influence of some of the early missionary familes was strong and the political influence stronger. During the later years of the nineteenth century there was some Anglo-American rivalry in the Hawaiian Islands, which might at the time as easily have become British as American. They became American in due course, as was perhaps inevitable: but the Union Jack is still the principal emblem on the State flag of Hawaii – at the hoist, right beside

* The family papers of the Ledyards, now at the University of Syracuse, indicate that the corporal was not highly thought of by anyone. Apart from having been once with Cook, his chief claim to fame seems to be the fact that he was probably the first American citizen deported from Russia for 'spying'.

the mast. This is the Hawaiian flag today. And that, one thinks, is the true Hawaiian testimony to the memory of James Cook.

To the average Englishman, the Cook discoveries and explorations were so remote as to belong almost to another world. Before him, even the virile and adventurous Americans knew nothing of their own west coast and had sent no ships to the Pacific. Contemporary reaction to the Cook voyages – apart from the few experts and far-sighted statesmen who could foresee their value – was slight, almost indifferent. It was all expressed far too well by the snap-jawed (and sometimes snap-judging) Samuel Johnson, who impatiently brought the loquacious Boswell to a halt whenever he mentioned Captain Cook. Johnson could see little use in voyages, or in books about them. The impressionable Boswell, after meeting Cook back from his second voyage, told Johnson that he 'felt a strong inclination to go with him on his next voyage'. (It was a pity he didn't: we certainly would know a great deal more about the Captain. And it was a better occupation than getting the pox round London.)

'Why, Sir, a man does feel so,' replied the oracle, 'till he considers how very little he can learn from such voyages.'

When the three volumes of Cook's Voyages were published, Johnson prophesied that they would be eaten by rats and mice in the warehouse before anyone read them through.

'There can be little entertainment in such books. One set of savages is very like another,' said he (as reported by Boswell).

Boswell: 'I do not think the people of Tahiti can be reckoned savages.'

Johnson: 'Don't cant in defence of savages.'

Boswell: 'They have the art of navigation.'

Johnson: 'A dog or cat can swim.'

Boswell: 'They carve very ingeniously.'

Johnson: 'A cat can scratch.'*

So, apparently, could the ancient lexicographer. It is fair to

* Boswell's *Life of Johnson*, edited by J. W. Croker. John Murray, London, 1860.

add that he knew Hawkesworth. Very likely, Boswell was being enthusiastic about the 'noble savage', an idea that Johnson saw right through.

It is very odd that we know so little about the personality, the human side, of Cook. He was a northerner of Scots and Yorkshire blood. He was brought up hard, of lowly status on a Yorkshire farm. He served years in Whitby colliers. He became, among many other things, an outstanding sailing-ship Master – an austere and lonely life both in the Royal and Merchant Navies. The qualities of character, judgement, and iron nerve necessary for success in that difficult and dangerous calling are not upon the surface, nor obvious at all, and exercised only in ships among seamen. I have had the good fortune to know four great sailing-ship captains. They were hard men, strong-faced, silent, with cold blue eyes – not at all readily approachable at sea, not all (I gathered) notable successes at home. They were what their calling made them. They dispensed with anecdote and they never 'yarned'. They wrote nothing of themselves and only what was essential of their ships. They could be met only on common ground, for they were hard men to know.

One imagines James Cook as one of these. He had the hard up-bringing, the competence, the qualities of leadership, the tremendous and greatly tried ability. The landsmen neither knew nor appreciated him, not even men of such discernment and ability as Banks and Solander. His seamen and his officers knew him. Many of them came back with him voyage after voyage, some unto death. They wrote little, though when he died the senior officers and a seaman or two paid their professional tributes. Banks was three years with the *Endeavour*, writing industriously much of the time: yet never once does he give more than odd hints on the human side of Cook, and very few of them. James Burney, the observant Fanny's brother, was years with Cook: he became a noted historian of Pacific voyages. Cook was at their father's house: he was seen about in London: he visited, on rare occasions, the homes of some of the great. His king knew him. Benjamin Franklin never saw him but admired him greatly. Diarists wrote of his voyages, of what he had done, of the allegedly noble savage in general and the Polynesian Omai in particular. Yet no Burney nor other diarist ever adds a word to increase our personal knowledge of the Captain. Perhaps his character was beyond them: but was he really as inscrutable as all that, a stern and cold-eyed image, his gaze for ever towards the far Pacific? Some called him affable. All called him strict. Many referred to his hot temper hastily roused by the slightest dereliction of duty, but just as quickly gone again, and forgotten. All referred to his extraordinary and all-embracing competence in whatever he undertook, his amazing ability to rise to do so much more than was asked of him, and, in the doing, carry the dignity, the leadership, the power of command as if he had been born with all these things and they developed naturally as he rose. Probably they did. He rose: but he never forgot those who stayed behind.

Yet we never know Cook as a person at all. We are entitled to wonder whether his contemporaries did.

There are several portraits of him done by good artists but these differ so much they could be of different persons. The Dance portrait in the National Maritime Museum at Greenwich shows a strong-faced, unworried seaman, fit, commanding, competent. The portrait in the Mitchell Library at Sydney (said to be by Hodges or Webber) is of a stern man, dour, withdrawn, forbidding, suffering from stomach ulcers. The few others fit between. Which if any of the artists got nearest to the man we have no means of knowing. The few statues of him were made long after his death: what manner of likenesses these may be no one who knew him ever had the chance to say. At least those in the Mall, near Admiralty in London, and on Whitby's west cliff looking over the harbour, look as one imagines Cook once did: that is as far as one may go.

His real Memorial is on the map of the world, above all the Pacific Ocean.

The truth was that Cook's worth could really be appreciated among his contemporaries by other seamen and the value of his achievements understood by a limited cognoscenti knowledgeable in the affairs of the sea, the story of discovery, and world geography. At a time of many spectacular fighting heroes, the seaman-explorer who disposed of comforting myths and found things exactly as they were, might expect little comprehension outside the services he had joined, both Royal and Merchant Navies, and the Royal Society which had accepted him. The public likes its drama to be obvious and its accepted heroes likewise. Cook was not an 'obvious' character to anyone; and the great English general public has no informed interest in its ships and seamen. It had not in the eighteenth century and it has not now.

It had then and still has the good fortune to be well served by splendid seamen, naval and mercantile, and able shipowners, builders, and insurers, whom it happily takes for granted and gives no thought except in time of war. For these are dedicated specialists, fortunately available in sufficient numbers when required. A great many come from Wales, Ireland, and Scotland. They serve and, for the most part, are not noticed. The myth of the island people with the sea in their blood, a race of 'born seamen', is just that: for seamen are trained, and not born. As for 'sea-mindedness', that is myth too. Living in islands in the same latitude as Labrador, on the lee side of the stormiest ocean in the world, at the receiving end for every dissipating hurricane and westerly storm, the vast majority is perpetually astonished at the arrival of bad weather and unable to see a coming squall. Sea traditions, and all that? It took them until the middle of the twentieth century to establish a National Maritime Museum and that was mainly achieved by four devoted men. They let Drake's *Golden Hind* rot and might well have left the *Victory* reach the same condition if it were not for Admiralty, the Society for Nautical Research, the Wylies, and Sir James Caird. They have heard of Columbus, Nelson, Drake. Sometimes it seems that the only sea 'tradition' which affects the great bulk of the heedless populace is acceptance of its seamen without thinking of thanking them, and God.

Against this background, James Cook remains an enigma. It is unlikely to be solved. Was he no more than the 'competent nonentity' who made professionally good? The seamen's seaman, capable and dedicated beyond the understanding of shore-dwellers? It is tantalizing that we know so little of him, despite the hundreds of thousands of words he wrote and the millions written about him: it was tragic that he was cut down in his prime. His seafaring was done, then: but he had shown an astounding capacity for development. With the long voyages all behind him, the Pacific still needed him. After him came the lawless and the profligate, the exploiters of whales and seals and sandalwood, the beachcombers, the unwilling colonists, the laggard seamen who took half-a-year and more to reach the young Sydney from England, with time-wasting, disease-spreading calls at Brazil on the way.

The new Australia missed Cook. So did the young New Zealand, for long regarded as mere sources for good spars for naval stores. Cook knew both these places better than anyone else on earth. He might have helped to get both countries started more effectively, and kept a useful and benevolent eye on Tahiti, the Hawaiian and the Friendly Islands as well.

This is surmise. The record stands. This is sufficient to establish James Cook, the farm-hand's son from Yorkshire who became Captain R.N. and gold medallist of the Royal Society, as the greatest explorer-seaman the world has known. The names of his brave ships may stand as his best epitaph – *Endeavour, Resolution, Discovery, Adventure*. To these must be added a fifth.

Its name is Humanity.

Technical Appendix

IN Cook's day the sort of small square-rigged ship he sailed was not at all complicated in the sense that the later-day big steel Cape Horners became. From open boat to closed hull, from a single mast with one sail to three masts and a bowsprit with two to four sails on a mast, the development was straight-forward and quite clear. First, in Viking times and the early Mediterranean, one mast sufficed to carry sail enough to give small vessels manageable way in anything of a breeze. The sail was stretched on a yard which hoisted on a pole, the mast. The mast was solid enough to stand without the support of 'standing' rigging: some light tackles were set up to windward as time went on, but the yard could be swung from the line of the keel – 'fore-and-aft' – to across the keel, when it is known as 'square'. The sail was 'set' from its simple yard by hauling down the corners (called clews), the clew to windward being the 'tack' hauled down by a line also called the tack, and the clew to leeward was 'sheeted' by a line called the sheet. When the ship sailed dead before the wind, both clews were held by sheets, which were secured on deck – 'belayed' to wooden pegs called belaying pins, or to pieces of wood conveniently nailed to the stanchions holding up the bulwarks (upper sides), called cleats.

Since one sail could be too large to handle easily as ships grew in size, the area of the sail was made changeable by converting the lower portion to a separate 'bonnet', which could be lashed on or left off. Ropes were rigged from the deck across the hoisted yard to run over the front of the sail down to the foot. These were 'buntlines', for their work was to 'bunt' the sail – stifle its folds – when it was required to take it in. Otherwise, as the yard was lowered, it would blow all over the men and obscure their vision. Clew-lines hauled up the clews. Halliards – the word comes from 'haul yard' – hoisted the yard. Later a small mast was added forward, right in the eyes of the ship, to become in time further divided into two – the bowsprit (a sprit at the bow), and the foremast, which used to be very far forward and in many ships took the place of the bowsprit itself (as in Chinese junks to this day). From the bowsprit (or on a yard slung from it), was set the 'spritsail' – the sail set on the sprit. This spritsail could be swung, like the old mainsail – called that because it was the main sail – from the line of the ship to right across it, and so it was a most useful manoeuvring sail. As it was so far forward, it would easily swing the ship even if she was standing still. It was therefore a most valued adjunct to the rudder.

The foremast carried one sail, at first, and this was the fore sail. There were limits to the sizes of these larger lower sails. To get more sail area, it was necessary to extend the height of the masts. Suitable trees were not abundant. Therefore a second mast, lighter than the lower, was erected on the last three or four feet of the lower-mast – 'stepped' there, as seamen say. This carried a lighter yard, which hoisted and lowered, and on this the sail was called the top sail. The support of a built-up mast like this required, in time, supporting lines fixed permanently from the masthead to the deck – from both the lower masthead and the topmasthead. This was the 'standing' rigging: it stood. There was already a fighting platform at the top of the mast (a useful spot for archers, and for throwing out boiling lead and so forth on assailants trying to capture the deck). This stayed, still called the 'top', even when a third lighter mast was added at the topmasthead further to increase the sail area. There was already a 'top'-mast so the third mast was known as the topgallant, and the sail set on it was the topgallant sail.

Both the fore and main masts now carried these three sails – the lower (called the course), the topsail, and the topgallantsail. When in light weather a small fourth sail was set above the topgallantsail, it was called the royal. A third mast was added towards the stern, behind the mainmast, because it was useful to set a sort of balancing sail there, usually a big fore-and-aft sail called the mizzen, or the lateen, as it was set as a lateen, such as Arab ships still use. A lateen is a sail set on a long pliable yard hoisted on a very strong short mast, and it has the advantages (in smaller ships) of adapting itself to swinging across the ship as a running sail, or along the fore-and-aft line (more or less) to 'ply', or beat. The wind coming into it gives the ship forward way. It is an aerofoil, no matter how primitive, and indeed a most interesting one in large Arab and Indian dhows. The square sail is more of a 'shover', a pusher, accepting the wind from its after side only. The lateen is much more like the sail the yachtsman uses, either as a gaff-and-boom or jibheaded mainsail, and this is how the lateen mizzen developed. The fore part of the lateen yard was sawn off and the after-part became the 'gaff', swinging on jaws behind – abaft – the mast. Sheets from its after corner restrained the lateen, now the 'spanker'.

Buntlines and clewlines took care of the square sails, taking their names from their own sails, and all led in orderly fashion to the same convenient places on deck so that the various gear was all grouped and could be found in an unlit ship during the darkest night. The spanker, like the spritsail, was a valuable manoeuvring sail. The big square sails did the heavy work of driving the ship and the spritsail and spanker balanced her, so that the steersman's work was simplified and the ship handled well. All these square sails were useful manoeuvring planes, too. Lines attached to the yard ends – the 'yardarms' – could swing the yards so that their sails could pull or push, brake or swing as the master willed. Sails on one mast could be braked ('backed') and on another filled, causing the ship to stay more or less in the same place: by manoeuvring the sails and rudder, the ship could be swung across the wind right into it, to bring the wind from one side quickly to the other. This was 'tacking'. In strong winds and high seas, the same effect could be achieved by running off before the wind, swinging round, and bringing the wind on the other beam. This was 'wearing'.

As standing rigging became stronger and stouter in the larger square-rigged ships, additional sails were set from it between the masts and before the foremast, called staysails and jibs. As bonnets were useless on the upper sails, these were 'reefed', that is, so contrived that a tuck could be taken in them from the yards aloft, and their area reduced in storms. One tuck grew in time to three. The tucks were secured by 'reef-points' which the sailors tied with reef-knots across the tucked-up sail. To do this the sailors stood on 'foot-ropes' attached below the yards, and they stretched the head of the reefed sail by means of reef 'earings' after first pulling up the sides of the sail with reef-tackles. It was windy when they reefed topsails: they needed tackles.

Sail area was reduced and sails taken in – furled – first by manipulating the controlling ropes which led to the deck. Sailors spoke of 'lines', not ropes: the number of items called 'ropes' was small even in the largest ships – the bell-rope, footrope, manrope, bolt-rope (round the sails), and one or two others. Every other item had a name such as the buntlines, clewlines, halliards, downhauls (to haul down the hoisting yards, and the heads of the jibs and staysails), sheets, tacks, gaskets (to wrap round the stowed sails), guys, shrouds, backstays (shrouds supported the sides of the mast, spread from the masthead to the ship's sides or beyond, outboard, to platforms there called 'channels': backstays were stays leading backwards, to support the topmasts and topgallantmasts without getting in the way of the swinging yards), bobstays (strengthening the bowsprit by hauling down on it, keeping it in position), gammonings and wooldings (lashings, the first on the bowsprit close by the bow, the second round the masts to help strengthen them).

It was all essentially simple. Everything had a name, a purpose defined and understood, and its own place. You learned these names and the functions of things quickly. Youths sent to sea rapidly became orderly persons, used to keeping things at sea in their places: the men saw to that, for the rope's-end swiftly followed unheeded word-of-mouth. Neither a sailing-ship nor her sailors could suffer fools gladly, or at all.

Technical illustrations are on the following pages.

SAILS AND RIGGING

A Jib
B Fore topmast staysail
C Spritsail
D Fore topgallantsail
E Fore topsail
F Foresail (or fore course)
G Main topgallant staysail
H Main topmast staysail
I Main topgallantsail
J Main topsail
K Mainsail (or main course)
L Mizzen topsail
M Spanker
N Fore-mast
O Main-mast
P Mizzen-mast
Q Jibboom (protruding beyond the bowsprit)
R Spritsail-yard
S Reef-points (used to gather up reefs, or tucks, in the sails)
T Braces (each yard has a pair, which lead to the deck)
U Bowlines (used to make the sails set better)

Braces and bowlines etc. take their precise names from the sails or yards they serve: e.g., the T on *the right* is the starboard fore topgallant brace. As the sails are set, it is also the lee fore topgallant brace, because it is on the lee side. The other side is then called the weather.

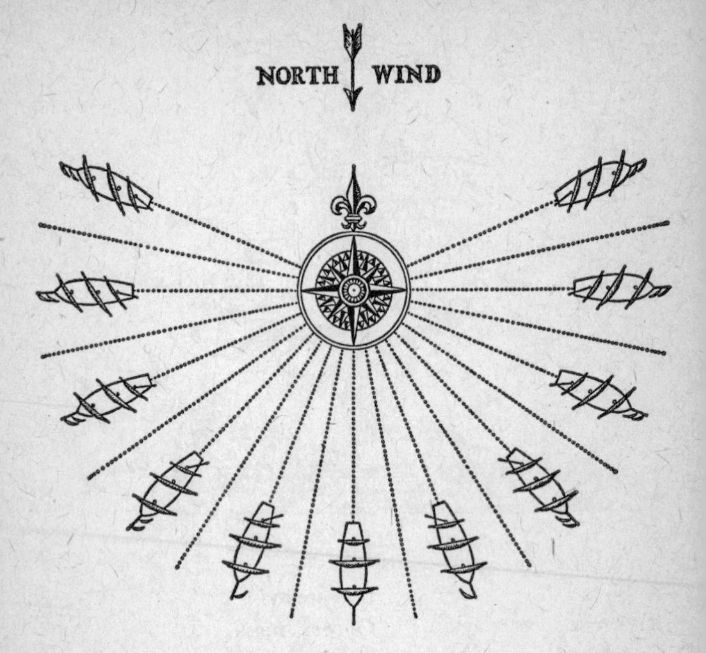

By swinging her yards (called bracing, because it was done by hauling and slacking braces, the tackles which swung the yards) a square-rigger altered the trim of her square sails. Here the wind is constant from North: with squared yards (right across the ship) the sailing-ship runs directly before it or with the wind just on either quarter, in the lower three sketches, gradually hauling up to the wind on either beam as the sketches ascend. The uppermost sketch on the right shows her as close to the wind's direction as she can trim her canvas to accept, and still give the ship headway. As the wind is on the port or left side, this is described as the port tack. A similar sketch on the opposite side shows the ship on the starboard tack, six compass points from the wind. This was about as close to the wind as the usual square-rigger could hope to sail. ... The wind should not be allowed to get in front of the sails, for this will take them 'aback' and the ship will not go ahead at all.

CAPSTAN

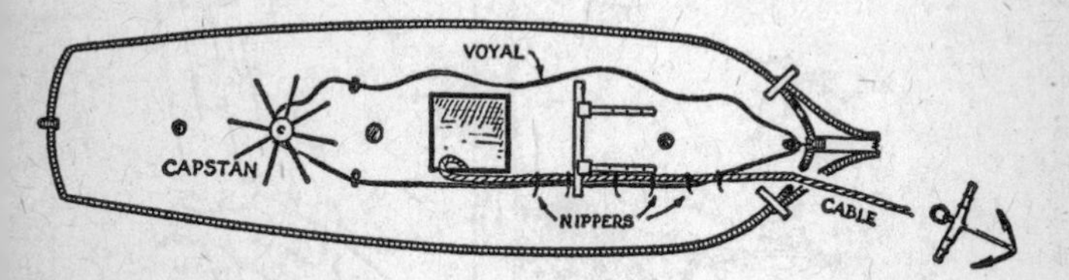

Anchor-work in Cook's ships was laborious and not very efficient. Hempen cables led inboard along the deck to the main capstan round which as many men as possible tramped while they hove away by capstan-bars. In larger ships they did not take the actual cable to the capstan but a lighter line secured by 'nippers' – a piece of braided cordage, easily removeable. This lighter line was called a 'voyal' or 'voyol', and it was used because the anchor cables were too big, wet, and awkward to go round the capstan. As it came inboard, the cable was stowed below out of the way, but taken up on deck in good weather later to dry out before any long period of stowage. The 'nippers' gave their name to smart lads who operated them. Both boys and the pieces of plaited cordage were, in time, called 'nippers'.

C
B
A

Sometimes square sails are thrown aback, or backed, to make some manoeuvre possible. Typical is that shown here, which is known as 'tacking' – throwing the ship round from one 'tack' to the other, which is essential when beating against a head (foul) wind. First, the ship sails along a little free from the wind, to have good way on her and so answer the helm and swing into the wind better. Then (figure A), the helm is swiftly put 'down' – into the wind – and everything possible done to assist the ship to swing into wind – the head-sheets eased off or let go, the spanker hauled to wind'ard, the fore tack and sheet let go, the mainsail hauled up. The ship now spins quickly into the wind's eye. As she swings (figure B) across the wind, all the main and mizzen yards are hauled round together very quickly, while the backed head-yards with their sails press the ship's head further and further round until the main and mizzen yards are *behind* the wind, and so the wind now begins to flow into and fill these sails. As they fill, and the ship begins again to gather headway, the foreyards are hauled round and the ship sails ahead on the other 'tack'. Such manoeuvres required nice judgement and good crews. Cook's ships handled well, because they were well sailed and well manned and also because of the placing of the masts, which exerted better leverage on such old ships than later-day full-rigged ships could manage.

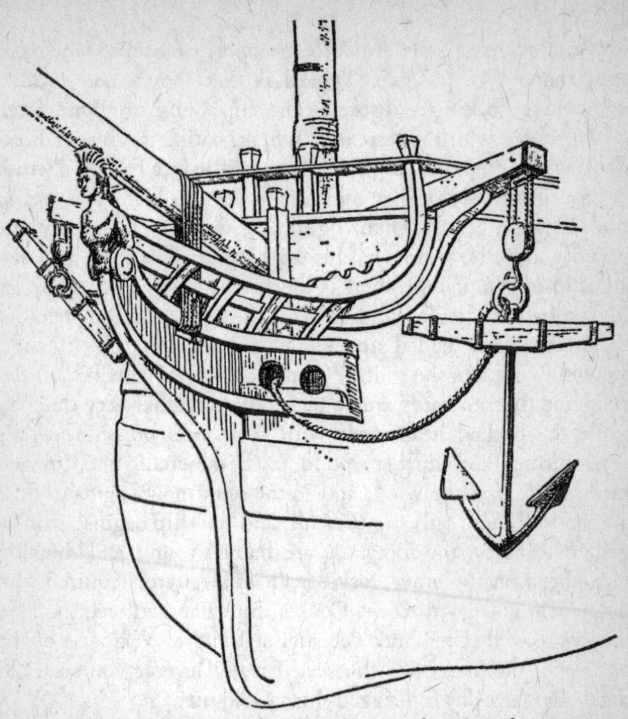

Bow of a ship such as Cook's *Resolution*, showing the cutwater, figurehead, bowsprit with base-lashing known as 'gammoning', cathead and cat-tackle for 'catting' the anchor. The stout baulk of timber which was the cathead carried sheaves in which the cat tackle was rove. It was often decorated with a carving of the head of a cat. The real purpose of the cathead was to keep the awkward big bower anchors clear of the ship's side while they were let go or hoisted up for stowage on passage. Here the starboard bower may be seen taken up. On long passages, the cat-tackles were unrove and the anchors lashed securely. Note wooden stock on anchors, business-sized flukes to grip the ground, hawseholes (with cable led through), bluffness of the old-type bows. One reason for this solidity was to give the ship 'lift' for'ard when she was plunging in a seaway, and so greater safety.

Shows a catted anchor being taken up to the ship's side by an additional tackle to the upper fluke. A chafing-board – additional heavy plank built on – protects the ship's side against damage caused by blows from the fluke when the ship is rolling. Note channel (platform to which the lower deadeyes are rigged), deadeyes (lower and upper, each with three holes called eyes, 'dead' because they have no sheaves and so are static), closed gunport.

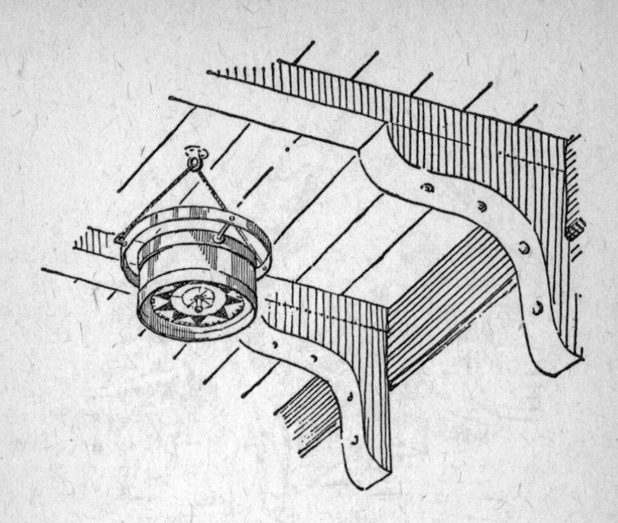

Cook's ships and indeed all long-voyage sailing-ships were fitted with a compass indicating the ship's course, which was hung above the captain's head. By this he could note what his ship was doing at any hour of the day or night. Whenever he wakened, there was the course recorded for him. The compass-card was face downwards.

Bibliography

IN any study of Captain Cook, first and foremost come the Hakluyt Society publications. Chief among these are the *Journals of Captain James Cook*, edited by Dr J. C. Beaglehole, C.M.G., Professor of British Commonwealth History at the Victoria University of Wellington, New Zealand, and others, published for the Hakluyt Society at the Cambridge University Press. By mid 1966, Volumes I and II, which cover Cook's first and second voyages, had appeared, together with a large folder of charts and views. The work of preparing Cook's own Journals for publication, almost two centuries after he began his *Endeavour* voyage, was immense and infinitely painstaking. These works do justice to the great seaman for the first time: the further volumes are awaited with interest. The Hakluyt Society's publications covering Byron's, Carteret's, and Quiros's Pacific voyages – Byron's *Journal of his Circumnavigation*, 1764–6, edited by Robert E. Gallagher; Carteret's *Voyage Round the World*, 1766–9, edited by Helen Wallis; *La Austrialia de Espiritu Santo*, edited by Celsus Kelly – are the standard works. Dr Beaglehole, who is the pre-eminent scholar in this Pacific field, has also given us the two volumes of *The Endeavour Journal of Joseph Banks*, 1768–71, published by the Trustees of the Public Library of New South Wales in association with Messrs Angus & Robertson, of Sydney and London; and the summarizing *Exploration of the Pacific*, third edition, in the A. & C. Black Pioneer Histories series, published in 1966.

With these works, one is likely to know as much about Cook in particular and the story of Pacific voyaging in general as one may need. But there are also Maurice Thierry's *Bougainville, Soldier and Sailor*; Dampier's *Voyages*, edited by John Masefield; the *Journal of Abel Janszoon Tasman and Franchoys Visscher*, published at The Hague in 1919; *The Voyage of Mr Jacob Roggeveen*, also from The Hague; and numerous collections of *Voyages*, of varying worth. I studied first Alexander Dalrymple's *Historical Collection of the Several Voyages and Discoveries in the South Pacific Ocean;* then James Burney, Charles de Brosses, and John Callander. Hawkesworth I looked into, but briefly: I spent time enough reading through him during my voyage in the ship *Joseph Conrad*. For the early Portuguese voyages, I read again carefully through Antonio Galvão's *Discoveries of a World* (a Hakluyt reprint of 1862), and R. H. Major's *Early Voyages to Terra Australis* with supplement *On the Discovery of Australia by the Portuguese* (Hakluyt Society, 1859), as well as Professor Edgar Prestage's *Portuguese Pioneers* (London, 1930) and several papers and works of his own which Dr Joaquim Bensaude had given to me. I also read Professor J. N. L. Baker's *History of Geographical Discovery and Exploration* (London, 1931).

Other books which have proved generally useful include Thomas Dunbabin's *The Making of Australasia* (London, 1922); *The Discovery of Australia*, by G. Arnold Wood (London, 1922); *Cook and the Opening of the Pacific*, by Dr J. A. Williamson (London, 1946); C. R. Beazley's *Dawn of Modern Geography* and Dr R. W. Giblin's *Early History of Tasmania* (London, 1928).

The standard biographies of Cook have been on my shelves for years, from Kippis (London, 1788) and Walter Besant (London, 1890) on – Kitson (London, 1907); Surgeon Rear-Admiral John Muir and Hugh Carrington (both published in London in 1939); Maurice Thierry, translated from the French and a little fanciful; Sir Joseph Carruthers, who takes the Hawaiians to task (London, John Murray, 1930), to Professor Christopher Lloyd (London, Faber & Faber, 1952). For contemporary details of Cook's third voyage, I have a selection of two-volume sets based on the journals left by Cook, Clerke, and Gore.

On the seamanship side the works are legion. Falconer and Steele make a sound beginning in a general way, but all sorts of small contemporary books help to throw a light on sea practices of Cook's day, so long as one reads them with something of a knowledgeable eye. Little things like Edward Ward's forthright *The Wooden World Dissected* etc. etc. (my copy, Edinburgh, 1752) and J. Cowley's more staid *Sailor's Companion and Merchantman's Convoy etc. etc.* (London, at the Globe in Paternoster Row, 1740) are most revealing. The North Sea sailor's life is well described in the *Memoirs of Henry Taylor*, from 1750 to 1821 (printed for its author in North Shields in 1821), for Taylor was a master in coal cats and was a Whitby seaman. A slim book called *Hints on Sea Risks* by Lieut. Jennings, R.N. (London, 1843); the report of the Select Committee of the House of Commons on Shipwrecks presented on 15 August 1836; the old Sailors' Pocket Books and Sea Officers' Manuals; John Masefield's *Sea Life in Nelson's Time* (London, 1905 to 1937), and Professor

Michael Lewis's naval historical works give an insight into conditions on the lower deck and in the great cabin.

I have a considerable library of these works, and others. For what I have not myself, I go to the library of the National Maritime Museum at Greenwich (and here I can consult original Cook's papers, read one of his Journals, study drafts of the *Resolution* and *Endeavour* and a model of the latter, look at the Dance portrait, examine paintings brought back by the various voyage artists). In Sydney, I go to the Mitchell Library, which has its own treasures: in Canberra, I call on my friend Mr H. L. White, C.B.E., the National Librarian, who has some treasures there, too: in Wellington, N.Z., I browse in the Museum, and call on Dr Beaglehole: in Whitby, on the Yorkshire coast, I go to the Museum, and wander through the streets of the little seaport town, and call in at Cook's house (the ship owning Walkers' really, where he lived when he was one of their apprentices) to visit the attic where he studied by candle-light in his teens.

As for the ships themselves, my friend Mr Marcus Fletcher, the former manager of the Whitby Shipbuilding Company which had a yard close to the spot where the old ship was built, has been making a study in detail of the *Endeavour* for years, to produce full plans and specifications from which it was hoped to build a replica, to be sailed to Sydney and established in the harbour there as a national maritime museum and memorial to Cook. Original drafts are still in the National Maritime Museum, with detailed measurements of the masts and spars, and these were used.

The Replica Endeavour *Plan*

THE replica *Endeavour* plan, first put forward in a paper which I read before the Council of the Royal Australian Historical Society in July 1962, came to nothing, although at first it seemed promising. The Whitby shipyard had to give up business, but another yard was found at Bideford, in Devon, which was quite capable of building such a ship, and interested. Marcus Fletcher died, but not before he had completed his specifications and plans. Sufficient shipwrights, still skilled in the art of handling timbers for the construction of 400-ton ships, and sailmakers, riggers, marine blacksmiths, and caulkers, were available in Devon, and some also in Yorkshire, Scotland, and Wales. A nucleus of experienced Cape Horn sailing-ship officers and seamen was available – for the last time – to form the basis of an excellent crew. Volunteers were legion. Good oaks were found.

An H.M.S. *Endeavour* Trust was formed in Australia (under the presidency of Vice-Admiral Sir John Collins, K.B.E., C.B., R.A.N., with Mr R. A. Dickson, vice-commodore of the Royal Sydney Yacht Squadron, chairman of the executive committee) and in the United Kingdom, where Viscount Boyd of Merton, P.C., C.H., was president and Admiral Sir Charles Madden, Bart, G.C.B., R.N., the executive director. Prominent shipping companies in the Australian and New Zealand trades, representatives of the Australian High Commission in England, the timber trade, the Hakluyt Society, the National Maritime Museum and the Society for Nautical Research were members.

At first, there appeared to be considerable interest. The Australian Federal Government and the governments of the States of New South Wales, Queensland and Tasmania either gave or promised some financial support, the Federal Government giving £A10,000. There was also a committee in New Zealand which planned to organize support, for the ship was to spend some months in New Zealand.

But the cost of building the bare hull alone, in the mid 1960s, had risen to more than £200,000. Sails, rigging, insurance, crew, stores and provisions for a year's voyage would have cost the best part of another £100,000. By 1 January 1967 – the latest date at which the keel could be laid if the ship were to be ready in time for working-up and then making the very long (for her) voyage – less than a third of this sum was in sight, nor did there appear any prospect then that more would be raised. To build the ship at all, a guarantee of her building costs was required. This could not be forthcoming and the project had, very reluctantly, to be given up.

The Three Voyages of Captain Cook

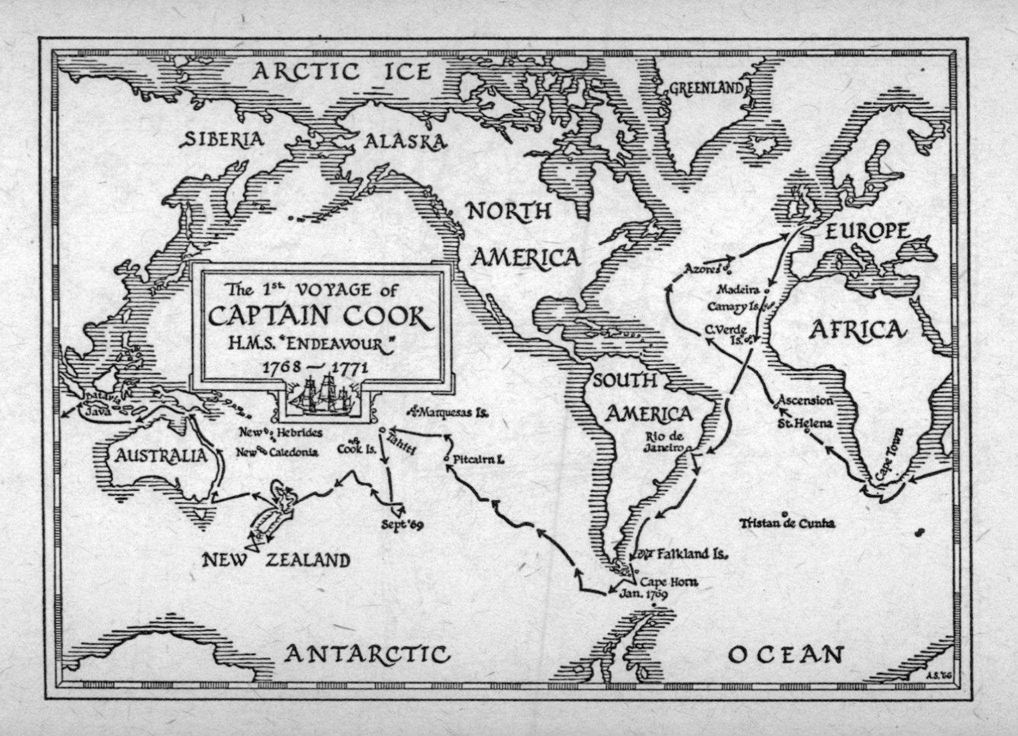
The 1st. VOYAGE of
CAPTAIN COOK
H.M.S. "ENDEAVOUR"
1768 — 1771
ARCTIC ICE
GREENLAND
SIBERIA
ALASKA
NORTH
AMERICA
EUROPE
AFRICA
Azores
Madeira
Canary Is.
C. Verde Is.
SOUTH
AMERICA
Rio de Janeiro
Ascension
St. Helena
Cape Town
Tristan de Cunha
Falkland Is.
Cape Horn
Jan. 1769
Marquesas Is.
Tahiti
Pitcairn I.
Cook Is.
Sept '69
New Hebrides
New Caledonia
Batavia
Java
AUSTRALIA
NEW ZEALAND
ANTARCTIC
OCEAN
A.S.'66

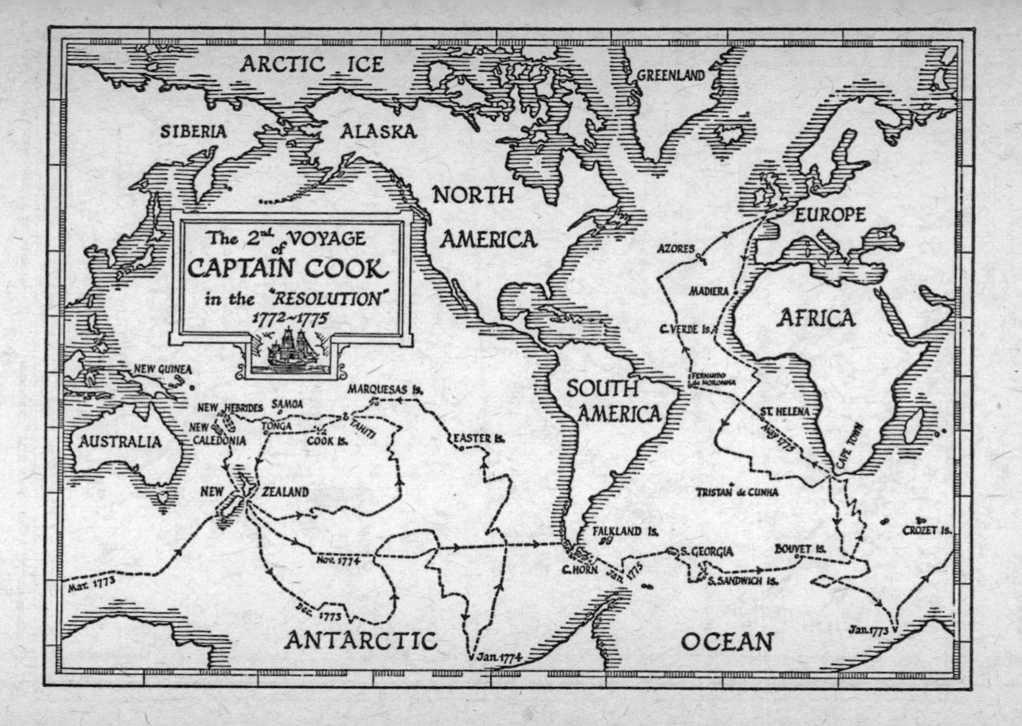
The 2nd of VOYAGE
CAPTAIN COOK
in the "RESOLUTION"
1772~1775
ARCTIC ICE
GREENLAND
SIBERIA
ALASKA
NORTH
AMERICA
EUROPE
AZORES
MADIERA
AFRICA
C. VERDE Is.
NEW GUINEA
FERNANDO de NORONHA
SOUTH
AMERICA
MARQUESAS Is.
NEW HEBRIDES
SAMOA
ST. HELENA
NEW
CALEDONIA
TONGA
TAHITI
AUSTRALIA
COOK Is.
EASTER Is.
May 1775
CAPE TOWN
NEW
ZEALAND
TRISTAN de CUNHA
FALKLAND Is.
CROZET Is.
S. GEORGIA
BOUVET Is.
Nov. 1774
C. HORN
Jan. 1775
S. SANDWICH Is.
Mar. 1773
Dec. 1773
Jan. 1773
ANTARCTIC
OCEAN
Jan. 1774

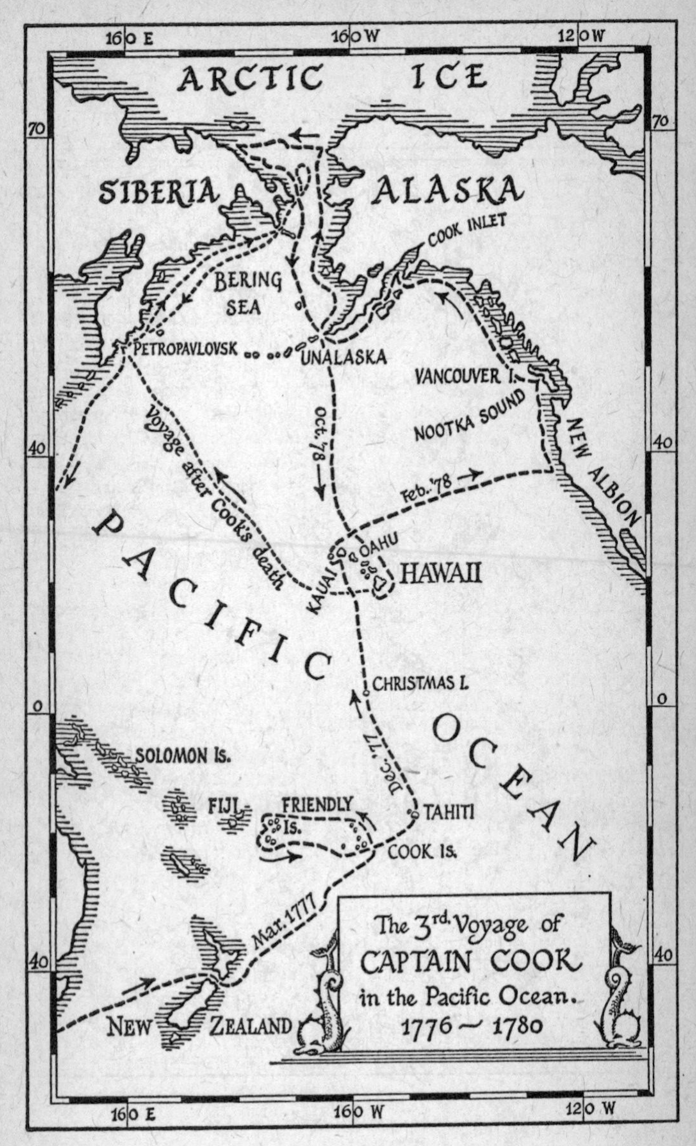
160 E
160 W
120 W
ARCTIC ICE
70
SIBERIA
ALASKA
COOK INLET
BERING SEA
PETROPAVLOVSK
UNALASKA
VANCOUVER I.
NOOTKA SOUND
NEW ALBION
40
Voyage after Cook's death
Oct. '78
Feb. '78
PACIFIC OCEAN
OAHU
KAUAI
HAWAII
CHRISTMAS I.
0
SOLOMON IS.
FIJI
FRIENDLY IS.
TAHITI
Dec. '77
COOK IS.
Mar. 1777
NEW ZEALAND
The 3rd. Voyage of
CAPTAIN COOK
in the Pacific Ocean.
1776 ~ 1780

Index